Plastic Problem

Editor: Danielle Lobban

Volume 440

First published by Independence Educational Publishers

The Studio, High Green

Great Shelford

Cambridge CB22 5EG

England

© Independence 2024

Copyright

This book is sold subject to the condition that it shall not,
by way of trade or otherwise, be lent, resold, hired out or otherwise
circulated in any form of binding or cover other than that in which it
is published without the publisher's prior consent.

Photocopy licence

The material in this book is protected by copyright. However, the
purchaser is free to make multiple copies of particular articles for instructional
purposes for immediate use within the purchasing institution.
Making copies of the entire book is not permitted.

ISBN-13: 978 1 86168 900 9

Printed in Great Britain

Zenith Print Group

Acknowledgements

The publisher is grateful for permission to reproduce the material in this book. While every care has been taken to trace and acknowledge copyright, the publisher tenders its apology for any accidental infringement or where copyright has proved untraceable. The publisher would be pleased to come to a suitable arrangement in any such case with the rightful owner.

The material reproduced in **issues** books is provided as an educational resource only. The views, opinions and information contained within reprinted material in **issues** books do not necessarily represent those of Independence Educational Publishers and its employees.

Images

Cover image courtesy of iStock. All other images courtesy of Freepik, Pixabay, Pexels and Unsplash.

Additional acknowledgements

With thanks to the Independence team: Tracy Biram, Klaudia Sommer and Jackie Staines.

Danielle Lobban

Cambridge, May 2024

Contents

Chapter 1: The Problem with Plastic

What is the problem with plastic?	1
Top 10 countries producing most plastic waste	2
Plastic facts: how long does it really take for things to break down?	6
Different types of plastic – recycling awareness	8
How bad is single-use plastic?	9
The myths surrounding single-use plastic	10
I avoided single-use plastic for a week – it should not be this hard or expensive	12
What are microplastics and why are they a problem?	14
'They're in the air, drinking water, dust, food …' How to reduce your exposure to microplastics	16

Chapter 2: Plastic and the Environment

Myths about plastic pollution are leading to public confusion: here's why	18
'Deadly' marine plastic hotspots revealed	20
Guide to the effects of plastic pollution in the ocean	22
Pandemic face masks could harm wildlife for years to come	26
New study into impact of plastic pollution on livestock and donkeys	28
Littering: the effects on wildlife	29

Chapter 3: Solving the Plastic Problem

15 ways to reduce your plastic use	30
Bioplastics: pros & cons and are they the future?	32
£3.2 million for innovation in plastics reduction	34
'Plastic-eating' enzymes to be deployed to combat waste polyester clothing	37
A sustainable future: unlocking plastic recycling with table salt	38
Chemical recycling: a game-changer for the plastic waste crisis?	39

Useful Websites	42
Glossary	43
Index	44

Introduction

Plastic Problem is Volume 440 in the **issues** series. The aim of the series is to offer current, diverse information about important issues in our world, from a UK perspective.

About *Plastic Problem*

In October 2023 a ban on some single-use plastic items was introduced in England. This book looks at different types of plastic, plastic pollution and the progress that is being made to tackle the problem. It also explores the pros and cons of plastic alternatives and considers if living 'plastic-free' is a possibility in the future.

Our sources

Titles in the **issues** series are designed to function as educational resource books, providing a balanced overview of a specific subject.

The information in our books is comprised of facts, articles and opinions from many different sources, including:

- Newspaper reports and opinion pieces
- Website factsheets
- Magazine and journal articles
- Statistics and surveys
- Government reports
- Literature from special interest groups.

A note on critical evaluation

Because the information reprinted here is from a number of different sources, readers should bear in mind the origin of the text and whether the source is likely to have a particular bias when presenting information (or when conducting their research). It is hoped that, as you read about the many aspects of the issues explored in this book, you will critically evaluate the information presented.

It is important that you decide whether you are being presented with facts or opinions. Does the writer give a biased or unbiased report? If an opinion is being expressed, do you agree with the writer? Is there potential bias to the 'facts' or statistics behind an article?

Activities

Throughout this book, you will find a selection of assignments and activities designed to help you engage with the articles you have been reading and to explore your own opinions. Some tasks will take longer than others and there is a mixture of design, writing and research-based activities that you can complete alone or in a group.

Further research

At the end of each article we have listed its source and a website that you can visit if you would like to conduct your own research. Please remember to critically evaluate any sources that you consult and consider whether the information you are viewing is accurate and unbiased.

Issues Online

The **issues** series of books is complemented by our online resource, issuesonline.co.uk

On the Issues Online website you will find a wealth of information, covering over 70 topics, to support the PSHE and RSE curriculum.

Why Issues Online?

Researching a topic? Issues Online is the best place to start for...

Librarians

Issues Online is an essential tool for librarians: feel confident you are signposting safe, reliable, user-friendly online resources to students and teaching staff alike. We provide multi-user concurrent access, so no waiting around for another student to finish with a resource. Issues Online also provides FREE downloadable posters for your shelf/wall/table displays.

Teachers

Issues Online is an ideal resource for lesson planning, inspiring lively debate in class and setting lessons and homework tasks.

Our accessible, engaging content helps deepen students' knowledge, promotes critical thinking and develops independent learning skills.

Issues Online saves precious preparation time. We wade through the wealth of material on the internet to filter the best quality, most relevant and up-to-date information you need to start exploring a topic.

Our carefully selected, balanced content presents an overview and insight into each topic from a variety of sources and viewpoints.

Students

Issues Online is designed to support your studies in a broad range of topics, particularly social issues relevant to young people today.

There are thousands of articles, statistics and infographs instantly available to help you with research and assignments.

With 24/7 access using the powerful Algolia search system, you can find relevant information quickly, easily and safely anytime from your laptop, tablet or smartphone, in class or at home.

Visit issuesonline.co.uk to find out more!

Chapter 1: The Problem With Plastic

What is the problem with plastic?

Plastic is everywhere around us, from the water bottles we drink out of, to the packaging of our favourite snacks, to the containers we use for storage. It has become an integral part of our modern lives, but what many of us may not realize is the growing problem associated with plastic. In this book, we will explore what plastic is, how it is made, and the various ways it is used, while also shedding light on the environmental challenges it poses.

What is plastic?

Plastic is a synthetic material made from polymers, which are long chains of molecules. These polymers can be moulded into different shapes and forms. As a raw material, plastic is a solid, but when heated and processed, it becomes a malleable substance that can be easily shaped. Plastic is incredibly versatile and is used in a myriad of products, from single-use items like straws and cutlery to durable objects like cars and electronics.

How is plastic made?

The production of plastic starts with the extraction of raw materials, such as oil or natural gas. These materials are then refined and transformed into ethylene or propylene, which are the building blocks of most plastics. Through a process called polymerization, these building blocks are chemically bonded, forming long chains of polymers, resulting in plastic.

Different types of plastic can be created by altering the ingredients and the manufacturing process. For example, polyethylene is a common type of plastic used in grocery bags and milk jugs, while polyvinyl chloride (PVC) is often used in pipes and building materials.

The many uses of plastic

Plastic has become ingrained in our daily lives and is used in numerous industries due to its versatility and durability. Here are some common applications of plastic:

- **Packaging:** Plastic is commonly used in packaging products such as food containers, bottles, and cling wrap. It helps to protect and preserve the contents, extending their shelf life.

- **Construction:** Plastic is used extensively in the construction industry for pipes, insulation, flooring, and roofing materials. It provides durability and insulation properties while being lightweight.

- **Transportation:** Plastics are vital in the production of automotive components, including dashboards, seats, and bumpers. Their lightweight nature helps improve fuel efficiency.

- **Healthcare:** Plastic plays a crucial role in the healthcare industry, providing safe and sterile materials for medical devices, such as syringes, IV tubes, and surgical instruments.

- **Electronics:** From computer casings to smartphone components, plastic is a key material in the electronics industry. Its electrical insulation and lightweight properties make it ideal for this purpose.

The problem with plastic

While plastic has numerous useful applications, its production, consumption, and disposal have raised significant environmental concerns. Here are some of the problems associated with plastic:

Pollution: Plastic waste takes hundreds of years to decompose, and much of it ends up in our oceans, rivers, and landfills. This pollution harms marine life and ecosystems, as animals can ingest or become entangled in plastic debris, leading to injury or death.

Microplastics: Plastic never truly disappears; instead, it breaks down into smaller pieces called microplastics. These tiny particles can be found in water bodies and even in the food we consume, posing potential risks to human health.

Carbon Footprint: The production of plastic involves the extraction and refinement of fossil fuels, contributing to greenhouse gas emissions and climate change. Additionally, the energy required to manufacture plastic further exacerbates its environmental impact.

Single-Use Culture: Plastic is often used for disposable items, such as packaging and utensils, contributing to a culture of convenience and excessive waste. Reducing single-use plastic consumption is crucial for minimizing its environmental footprint.

Top 10 countries producing most plastic waste

By Inemesit Ukpanah

The growing problem of plastic waste pollution severely impacts our environment. Plastic waste is produced yearly in landfills, oceans, and natural areas. To remedy this problem, global action is urgently required. Over the years, 9.2 million and 12.7 million tonnes of plastic have been generated, with one waste truck's worth of plastic waste in the ocean every minute. If they do biodegrade, it may take hundreds of years.

According to UN Environment Programme research (UNEP), if current trends continue by 2050, there will be fewer fish due to plastic waste. On an annual level, plastic waste generated per person averages 221 kg in the United States and 114 kg in European Organisation for Economic Co-operation and Development (OECD) countries meanwhile, Japan and Korea generated 69 kg on average.

The highest total plastic waste per capita

Despite our efforts to minimise plastic consumption, ten countries stand out for generating the most plastic waste globally. Some countries have a considerable responsibility to combat the ongoing plastic waste crisis that threatens the health of our planet.

1. United States of America

Plastic use has increased significantly in the United States over the previous four decades, with at least 85% of municipal plastic waste disposed of in landfills. The United States produces over 42 million metric tonnes of plastic waste each year, which equates to approximately 130 kg of rubbish for every American. The recycling rate for plastic waste in the United States is declining, with only 5–6% recycled as of 2021 statistics. The infrastructure has failed to keep pace with the massive growth in American plastic production. This results in mismanagement, incineration, and inefficient waste disposal in landfills, causing up to 2.2 million tonnes of plastic to 'leak' into the environment.

2. India

India is one of the world's most significant contributors to plastic waste, generating about 9.46 million tonnes annually. The country is currently facing a severe challenge with plastic waste. The amount of plastic waste generated in the country is staggering and is becoming a growing environmental concern. The root cause of this issue is the need for proper waste management infrastructure in India, which leads to littered streets, overflowing landfills, and plastic waste infiltrating water bodies.

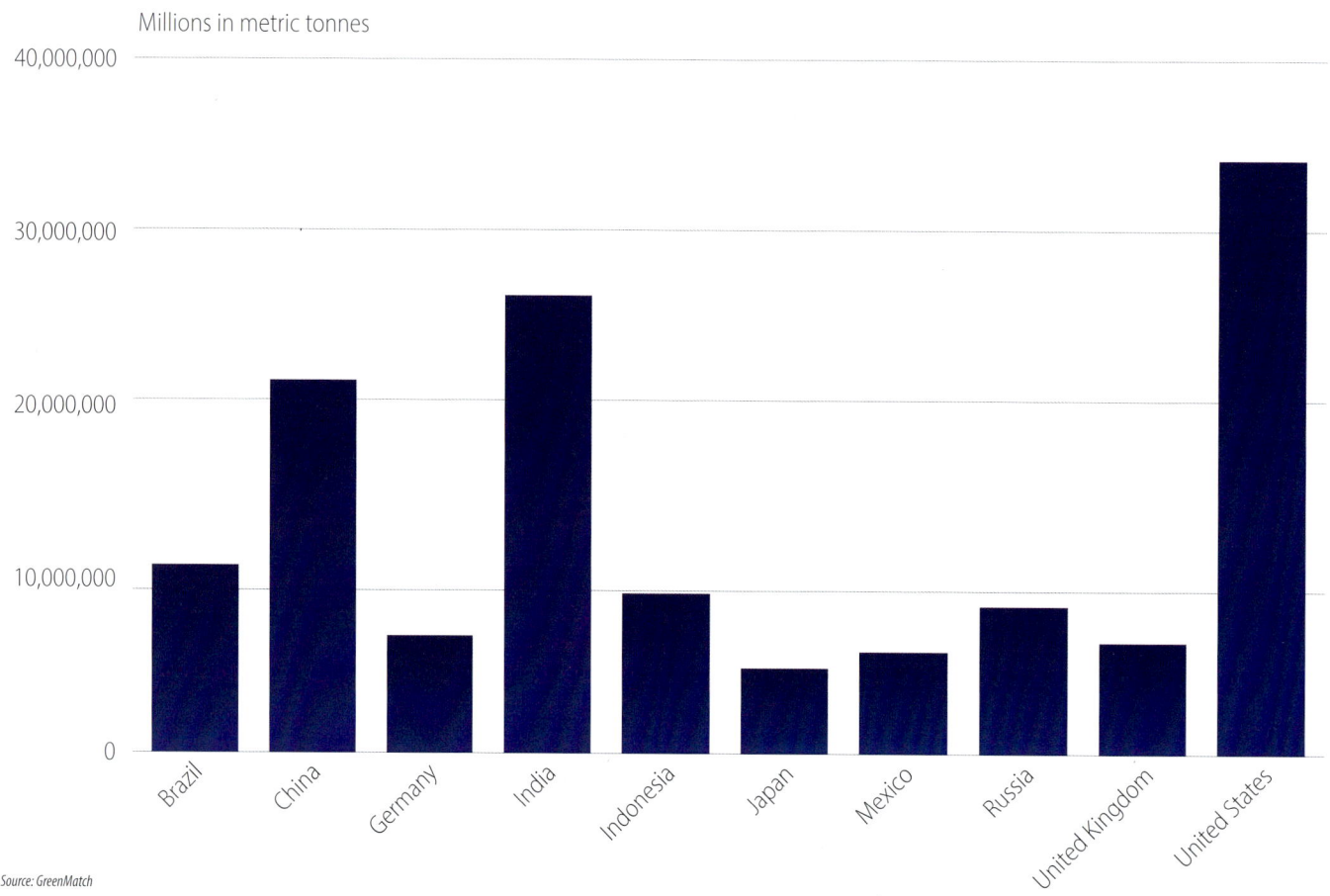

Top 10 countries producing the most plastic waste

Source: GreenMatch

With the negative effect of massive tonnes of plastic waste annually, of which 40% goes uncollected, the Indian government is fighting back by introducing an initiative, the Plastic Waste Management Rules. These programmes aim to reduce the use of single-use plastic, increase recycling, and promote proper waste management while raising awareness about the issue.

3. China

China is the biggest producer of plastic, with about 60 million tonnes of plastic waste, yet only 16 million tonnes were recycled. However, the country has taken steps to combat plastic waste pollution, including a ban on single-use plastics and a focus on the circular economy. With as much as 70.6% incinerated, recycled, and mismanaged, triggering environmental problems, per capita plastic consumption is well behind the United States.

4. Brazil

The plastic waste problem in Brazil is a pressing issue that demands urgent attention. The world's fourth largest producer of plastic waste, behind only the USA, China, and India, produces about 11.3 million tonnes annually, of which only 1.28% is recycled. This means that a significant amount of plastic waste ends up in landfills, mismanaged or in the ocean, posing a severe threat to the environment and marine life. Of Brazil's single-use plastic items each year, 13% are products such as plates, glasses, cutlery, plastic bags, and straws, contributing to Brazil's plastic waste problem.

5. Japan

It is concerning that Japan, a country known for its environmental awareness and cleanliness, is facing a severe plastic waste problem. Japan produces around 9 million tonnes of plastic waste annually, with more than 40% being disposable plastic, such as packaging and food containers. Despite being one of the largest consumers of plastic packaging globally, much of this waste ends up in landfills or the ocean, posing a significant threat to the ecosystems. The recycling rate of waste in Japan is 19.9%. Although the country has an efficient method of collecting recyclable materials, a significant amount of plastic waste is either incinerated or exported to other countries for processing.

6. Indonesia

Indonesia generates approximately 7.8 million tonnes of plastic waste annually, with 4.9 million tonnes being mismanaged, such as uncollected waste, disposed of in open dumpsites, or leaked from improperly managed landfills. Plastic waste has affected Indonesian rivers and the ocean, with more than 600,000 tonnes of plastic dumped, mostly from rural areas, due to an uncollected system. This has led to the Indonesian government launching a programme that will pay thousands of traditional fishers to collect plastic waste from the sea, aiming to reduce marine plastic waste by 70% by 2025.

7. Russia

Russia's plastic waste production is increasing, and it is one of the few countries whose plastic waste output is increasing, to 8.47 million tonnes. The country faces significant challenges with inadequate waste management systems, a lack of recycling infrastructure, and insufficient awareness. Although some recycling programmes are in place, they are not widely available or well-established.

8. Germany

Germany is one of Europe's biggest producers of packaging waste, especially plastics, despite being celebrated as a world leader in recycling. The country generates approximately 6.5 million tonnes of plastic waste pollution each year, with a considerable portion originating from packaging materials. According to Philipp Sommer, a circular economy expert, Germany recycles much less plastic packaging waste than commonly understood, with only 38% recycled plastic waste.

9. United Kingdom

The United Kingdom is one of the world's most significant sources of plastic waste per person contributing to approximately 6.4 million tonnes of plastic waste. Despite the government's best efforts, the country still generates considerable plastic waste, and the recycling infrastructure must keep pace with the significant growth. To address these challenges, the UK government has implemented several measures to reduce plastic waste, such as the plastic bag charge and the ban on microbeads, that ends up in landfills.

10. Mexico

Mexico is one of the countries that suffer from plastic pollution, with 5.9 million tonnes of plastic waste being a significant contributor to environmental degradation, with only 15% being recycled. Although Mexico has a high % percentage collection rate of 91%, most of this waste is improperly disposed of in unregulated landfills or enters the environment as litter. This improper disposal leads to incineration, which releases toxic chemicals into the air. The issue is particularly severe in Mexico City, the country's capital, which generates an estimated 13,000 tonnes of waste per day, up to 40% of which is plastic.

Although these nations generate substantial volumes of plastic waste statistics with minimal recycling, some are better at processing plastic waste safely. In contrast, middle-income and low-income countries still developing their infrastructure tend to produce more mismanaged waste plastic. The severity of the plastic pollution predicament plaguing portions of Asia, specifically India, China, and Indonesia, where mismanaged plastic refuse has reached critical levels, cannot be overstated.

The diverse methods by which countries deal with discarded plastics largely depend on a nation's economic status. This is because wealthier countries are generally able to either incinerate, recycle, or adequately landfill their plastic refuse. However, less affluent societies have no choice but to dispose of such detritus through uncontrolled dumping or open burning.

Although low-to-middle-income countries sadly struggle with mismanaged plastic waste pollution that ultimately flows into their waterways, seas, and ocean. Though intertwined and multi-faceted, the proliferating problem of plastic pollution demands collaborative and coordinated worldwide countermeasures.

Environmental impact of plastic waste

The adverse effects of plastic waste on the climate are numerous and far-reaching. Once hailed as a 'phenomenon material,' plastic has become one of the most burning environmental issues encyclopedically. This is our reality today, as plastic waste significantly wreaks havoc on our biodiversity, ecosystems, animal and plant life, and human health.

As plastic waste pollution requires immediate attention, we start by using reusable material rather than a disposable commodity that's quickly discarded. It is alarming to note that the global recycling rate for plastic waste is a mere 9%, while a staggering 22% needs to be adequately managed. This means using fewer materials to make plastic more easily recyclable or biodegradable. Increasing recycling facility availability will limit the number of polymers in manufacturing.

Carbon footprint reduction

Several countries have enforced successful actions and programmes to reduce plastic waste products and enhance disposal methods. Germany has joined a transnational coalition to tackle plastic waste. The 'Plastic Free July' program encourages individuals and businesses to reduce plastic consumption and waste in Australia.

Meanwhile, Norway has introduced a plastic bottle deposit system, resulting in a high recycling rate and reduced plastic waste. Plastic waste pollution is predicted to increase from 9% in 2019 to 17% in 2060, with more incineration and landfilling. Nevertheless, plastic waste is projected to increase utmost in non-OECD countries, driven by profitable growth in rising economies in Africa and Asia.

Global plastic waste

Recent statistics indicate that plastic waste is approximately **30% of all waste produced worldwide.**

The world generated **242 million tonnes of plastic waste,** 12% of all municipal solid waste.
The majority of this waste came from three regions:

- North America **35** million tonnes
- Europe **45** million tonnes
- East Asia & The Pacific **57** million tonnes

An estimated
81% of global ocean plastic waste
was emitted from Asia in 2019

Countries with a smaller geographical area, longer coastlines, high rainfall, and poor waste management systems are more likely to wash plastics into the sea.

Of the global plastic waste, it is estimated:

55% was discarded

25% was incinerated

20% was recycled

Source: GreenMatch

Projection of plastic waste in 2060

Plastic waste is expected to increase to **1,230 million metric tons produced by 2060**

Recycled

Recycling rates are projected to double from 9% (33 metric tonnes) in 2019 to 17% (176 metric tonnes) in 2060. This indicates a significant increase in recycled waste and a potential reduction in waste that ends up in landfills.

Incinerated

The proportion of waste that's been burned is expected to go up from 19% (67 metric tonnes) to 28% (179 metric tonnes) in 2060, while the relative contribution would remain steady over time.

Landfilled

Plastic waste sent to landfills would be tripled from 19% (17MT) in 2019 to 50% (507MT) in 2060. The levels of plastic waste being sent to landfills are set to increase significantly in the upcoming years, putting even greater pressure on scarce land resources, especially in areas near urban centres.

Mismanaged

With projections showing it to nearly double from 22% (79MT) in 2019 to 28% (153MT) in 2060, which would be driven mainly through Africa and Asia, if policies are not strengthened, mismanagement continues to escalate along with the increase in waste production.

Source: GreenMatch

Reducing plastic waste

To protect the environment, vigorous measures must be taken to limit plastic waste production and enhance disposal procedures. One efficient strategy to accomplish this is to promote reusable bags and containers, which have been shown to reduce the quantity of plastic waste generated drastically. Investing in improving recycling infrastructure and technology is also critical to ensuring that more plastic waste is recycled rather than ending up in landfills or the oceans.

Collaboration between governments and businesses can aid in developing policies and incentives that encourage using sustainable materials and practices. Adopting a holistic approach to reducing plastic waste will pave the road for our world's cleaner and healthier future.

Solutions

- **Recycling:** Collecting waste items and converting them into raw materials that can be utilised to create other valuable products
- **Incineration:** The burning of garbage to generate energy.
- **Landfilling:** The disposal of waste in designated areas.
- **Pyrolysis:** Heating waste without oxygen to produce fuel or other compounds.
- **Bioremediation:** Using microorganisms to break down plastic waste into non-hazardous chemicals.

The United States generates more than ten times the daily per capita plastic waste of countries like India. Plastic buried deep in landfills can leach harmful chemicals that spread into groundwater. The procurement and polymerisation of petrochemicals to produce plastics fuel the continual combustion of non-renewable fossil fuels. This has led to consequent emanation of greenhouse gases, intensifying the precarious perturbations in global temperatures and weather patterns.

Therefore, individuals can take action by reducing plastic use and participating in waste collection. Supporting policies on reduced plastic packaging is a significant step towards reducing the statistics on plastic waste. Let's work together to safeguard our environment and create a sustainable future for future generations.

26 July 2023

The above information is reprinted with kind permission from GreenMatch.
© 2024 GreenMatch

www.greenmatch.co.uk

Plastic facts: how long does it really take for things to break down?

We all know that plastic lasts for a long time. A lightweight, cheap, easily manufacturable material, plastic is a super material when it comes to making things. While the longevity of plastic can be seen as a benefit when making products, what happens to that plastic when we no longer need it?

Recycling is something we should all do more, but when items can't be, or aren't recycled, how long do they really take to breakdown?

Everyday items such as a paper towel, a newspaper, or a cardboard box, can take from 2–4 weeks up to 2 months. These paper-based items degrade quickly; however our other household waste isn't so easily disposed of.

Plastic shopping bags: 20 years

Usually only used for the short journey from the shop to the car, and then the car to your house, shopping bags only have a working life of a few minutes. This is horrifying when you know that these bags can take up to 20 years to break down!

Why not take the plastic bags you have at home and reuse them as much as possible, taking them to a recycling point when they eventually break. Make a switch to fabric tote bags instead for your shopping? Not only can these bags be used for years, they are washable, and at the end of their lives they can be repurposed, and recycled easily.

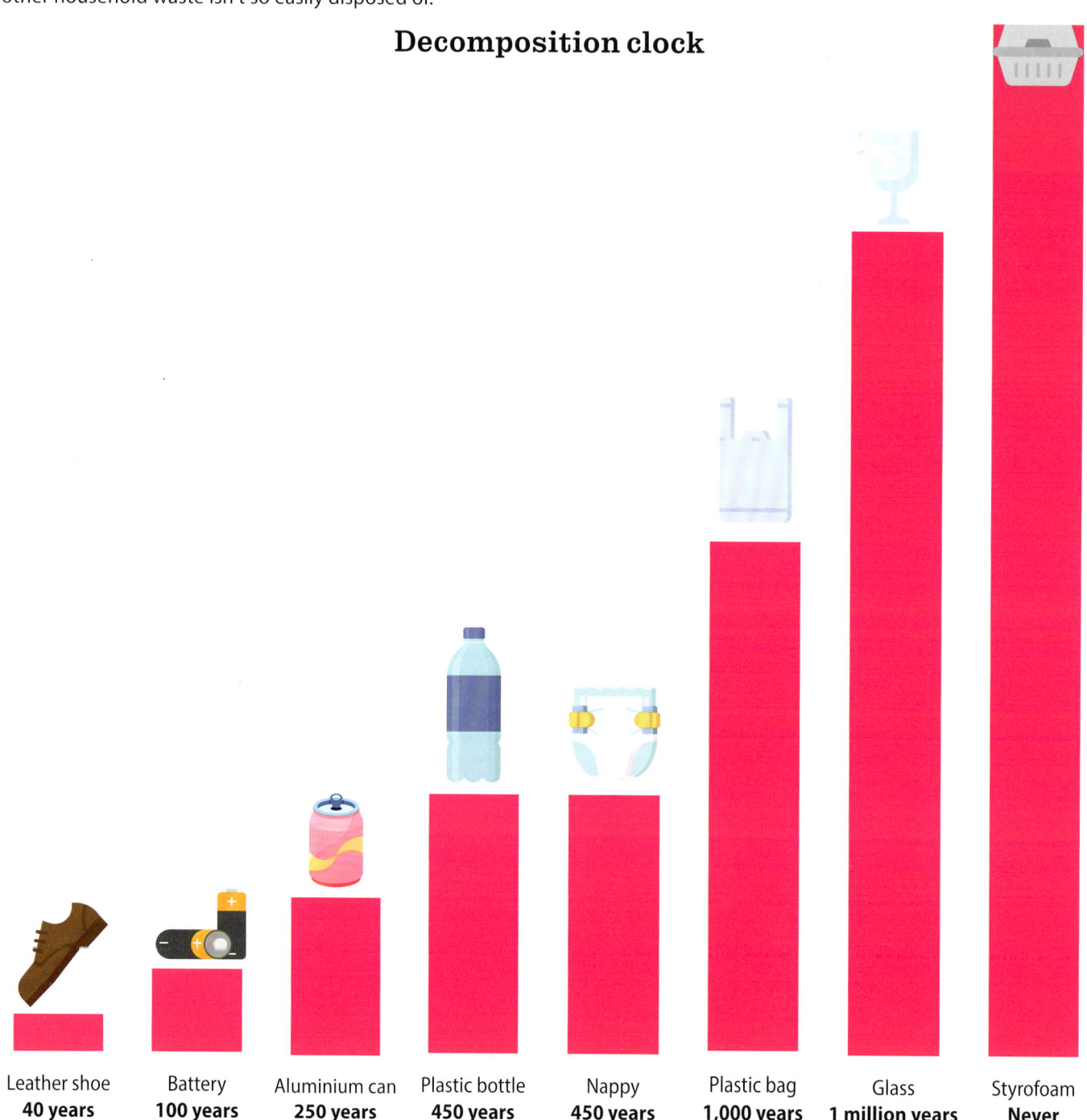

Decomposition clock

Leather shoe	Battery	Aluminium can	Plastic bottle	Nappy	Plastic bag	Glass	Styrofoam
40 years	100 years	250 years	450 years	450 years	1,000 years	1 million years	Never

Takeaway coffee cups: 30 years

Takeaway coffee cups, while made from paper, as lined with plastic to keep them waterproof. This means it cane take 30 years for each cup to breakdown. When we consider that 7 million coffee cups are thrown away in the UK every single day, we can quickly see how this is a massive problem!

Switching to a reusable cup is not only great for the planet, but your wallet also, as many coffee shops offer discounts when you bring your own cup.

Wet wipes: 100 years

Wet wipes cause around 93% of all sewer blockages, causing millions of pounds of damage a year in thte UK. These wipes are used in huge quantities and only have a useful life of a couple seconds.

The Marine Conservation Society released a fact that an average of 18 wet wipes were found per 100m of beach at their Great British Beach Clean 2020. A truly staggering amount of wipes to be found!

After being flushed, or thrown away, these wipes can take a century to break down. Made from plastic, they break down into microplastic particles, releasing chemicals as they degrade. A shocking 11 billion wet wipes are used in the UK every year, so the time is now for all of us to cut out wet wipes as much as humanly possible.

Aluminium cans: 250 years

Drink and food cans can take up to 250 years to break down. This is so frustrating to learn, when we know that aluminium cans are an infinitely recyclable material. Every minute of every day, an average of 113,200 aluminium cans are recycled across the globe. A recycled drinks can be recycled and back on supermarket shelves as new drink cans in as little as 60 days; and when we consider that it takes a quarter of a century for these cans to break down when they aren't recycled, it seems not only foolish but irresponsible not to recycle our aluminium cans.

Plastic bottles: 450 years

Plastic bottles are used for a multitude of liquids. From drinks, to cleaning chemicals, to shampoos, the list is endless. Easily recyclable, these bottles can be made into new items with ease, however the majority of plastic bottles are not recycled, being disposed of incorrectly.

Refilling a bottle, rather than buying new, can have a huge impact on reducing plastic waste. A refillable bottle has multiple benefits besides from being more environmentally friendly. On top of cutting down on plastic waste, it also saves you money as you do not need to buy a bottle each time you are thirsty.

> 'If just 1 in 10 Brits refilled just once a week, we'd save around 340 million plastic bottles a year.'

Plastic toothbrush: 500 years+

Our bathrooms are huge sources of plastic pollution, from microbeads in body washes, to wet wipes, to cotton buds, and sanitary products. An unlikely plastic problem, however, are our toothbrushes. There are 3.5 million toothbrushes sold annually worldwide. Made from nylon and polypropylene plastic, these toothbrushes often end up in landfill where they can take up to 500 years or more to decompose.

A bamboo alternative takes significantly less time to break down and therefore is a more sustainable option.

Conclusion

While these timelines are great for working out how long plastic items can last, what we aren't addressing is what happens after they finally break down? Unlike other materials, plastics do not biodegrade, they simply break into smaller and smaller pieces known as microplastics.

21 August 2023

The above information is reprinted with kind permission from Life's a Beach.
© 2024 Life's a Beach

www.lifesabeach.org

Different types of plastic – recycling awareness

The Resin Identification Code (RIC), commonly known as recycling numbers or plastic recycling codes, is a system used to identify the type of resin used in plastic products. The RIC consists of a number inside a triangular arrow symbol and is found on the bottom of plastic containers and products. Each number represents a specific type of plastic resin used in the item. Understanding these codes is important for proper recycling and waste management. Here are the different recycling numbers and what they mean:

PET or PETE (polyethylene terephthalate)

- **Description**: PET is a commonly used plastic for beverage bottles, food containers, and synthetic fibres.
- **Recycling number:** 1
- **Symbol:** Number 1 inside the recycling triangle.
- **Meaning:** PET plastics are recyclable and can be turned into a variety of products, including new bottles, polyester fibers, and packaging materials.

HDPE (high-density polyethylene)

- **Description:** HDPE is a sturdy plastic used for milk jugs, detergent bottles, and various containers.
- **Recycling number:** 2
- **Symbol:** Number 2 inside the recycling triangle.
- **Meaning**: HDPE plastics are recyclable and can be converted into new containers, pipes, and plastic lumber.

PVC (polyvinyl chloride)

- **Description:** PVC is a versatile plastic used in pipes, vinyl flooring, and certain packaging.
- **Recycling Number:** 3
- **Symbol:** Number 3 inside the recycling triangle.
- **Meaning:** PVC is less commonly recycled and can release toxic chemicals when incinerated. Proper disposal and recycling methods are crucial for PVC products.

LDPE (low-density polyethylene)

- **Description:** LDPE is a flexible plastic used for plastic bags, shrink wrap, and some containers.
- **Recycling number:** 4
- **Symbol:** Number 4 inside the recycling triangle.
- **Meaning:** LDPE plastics are recyclable, but the availability of recycling programmes for LDPE items varies. Some grocery stores accept plastic bags for recycling.

PP (polypropylene)

- **Description:** PP is a durable plastic used in bottle caps, food containers, and packaging.
- **Recycling number:** 5
- **Symbol:** Number 5 inside the recycling triangle.
- **Meaning:** PP plastics are recyclable and can be transformed into products like battery cases, brushes, and brooms.

PS (polystyrene)

- **Description:** PS is a lightweight plastic used in foam products, disposable cutlery, and packaging materials.
- **Recycling number:** 6
- **Symbol:** Number 6 inside the recycling triangle.
- **Meaning:** PS plastics are less commonly recycled and may be accepted by limited recycling programmes. Expanded polystyrene (EPS) foam is often not recyclable and poses environmental challenges.

Other plastics (miscellaneous)

- **Description:** This category includes various other plastic resins, such as polycarbonate (PC) and other specialty plastics.
- **Recycling number:** 7
- **Symbol:** Number 7 inside the recycling triangle, often accompanied by a specific resin code (e.g., 7 PC).
- **Meaning:** Number 7 plastics are a mix of different resins and may not be widely accepted by recycling facilities. Some number 7 plastics, like polycarbonate, contain BPA, a chemical of concern, and may have recycling limitations.

It's important for us all to be aware of these recycling numbers and their meanings to make informed decisions about recycling and waste disposal. Not all plastic items are recyclable in every area, so checking in with your waste carrier will help you make the right choice about your plastic waste.

9 November 2023

The above information is reprinted with kind permission from NWS.
© 2024 Nationwide Waste Services.

www.nationwidewasteservices.co.uk

How bad is single-use plastic?

From up in the mountains to down on the seabed, plastic waste is everywhere. If we're to end plastic pollution, the biggest contributors need to take responsibility and minimise the waste they're creating. Find out why single-use plastic is such a problem and how to hold polluters to account.

What is single-use plastic?

Single-use plastic is plastic we use once and throw away, often as quickly as minutes, such as disposable cups at parties. Half of the plastic produced each year is single use, which is nearly the same weight as the entire human population. A lot of single-use plastic ends up in our oceans, and by 2050, there will be more plastic in our ocean than fish.

The plastic problem

Plastic pollution is any plastic which ends up in the environment. It's incredibly harmful to all living things, as well as their habitats. It's important to remember that plastic sticks around in the environment for ages, threatening wildlife and spreading toxins.

As much as 12 million tonnes of plastic is poured into our oceans every year. Many animal species, including large mammals and tiny zooplankton, ingest plastic, mistaking it for food, causing some to die as a result.

But it's not only animals that suffer. Microplastics, plastics that are 5 mm or smaller, are found everywhere. They are created by bigger plastic objects breaking down, or are shed from synthetic clothes, carpets and vehicle tyres. It's estimated we could be consuming around 20 kg microplastics in our lifetime, from simply eating, breathing and drinking. This is really worrying because plastic products contain chemical additives, some of which have been linked to serious health issues like infertility and cancer.

More shockingly, up to 1 person every 30 seconds dies in a low-income country as a direct result of plastic waste disposal, cleaning up the mess created by mega corporations.

What's your plastic footprint?

It's understandable that we'd want to take individual action against the plastic waste we're generating. Here in the UK we generate 5 million tonnes of plastic waste each year, which equals roughly 75 kg per person. But with so many single-use plastic items around, it can feel incredibly difficult, impossible even, to avoid it altogether.

We challenged three contenders to give up single-use plastic and discovered just how difficult it is to give it up: 'it's insidious', said one of the contenders.

Who's responsible?

While our contenders were able to spend time planning their purchases and cover the cost of more expensive alternatives, a lot of people simply don't have that luxury. And some single-use plastic is essential, such as plastic straws to help ensure people with certain disabilities can drink safely and independently.

The problem with focusing purely on public consumption is that it shifts the blame to individuals rather than the real culprits: the major corporations producing single-use plastic products.

Single-use plastic league table

While some of us can try to reduce our own plastic consumption, there's no denying businesses and corporations are the most responsible for our plastic pollution crisis.

Common household names contribute significant plastic waste. The Break Free From Plastic movement conducted a brand audit of pieces of plastic waste found by volunteers across six continents. This involved 11,184 volunteers from across 45 countries collecting plastic waste for one week and recording the brand of the waste.

Volunteers recorded their findings, exposing some of the biggest plastic waste offenders. For three consecutive years, Coca-Cola took the not-so-coveted first place, producing more pollution than the next two top polluters combined.

Here are the top 10 plastic polluters:

1. **Coco-Cola:** 39 countries, 19,826 plastics
2. **Pepsico:** 35 countries, 8,213 plastics
3. **Unilever:** 30 countries, 6,079 plastics
4. **Nestlé:** 30 countries, 4,149 plastics
5. **Procter & Gamble:** 30 countries, 1,939 plastics
6. **Mondelez:** 28 countries, 2,065 plastics
7. **Philip Morris:** 26 countries, 1505 plastics
8. **Danone:** 25 countries, 3,223 plastics
9. **Mars:** 24 countries, 961 plastics
10. **Colgate-Palmolive:** 22 countries, 941 plastics

Together, Break Free From Plastic volunteers collected a massive 330,493 pieces of plastic waste and 58% of it was branded.

We need to hold the worst plastic polluters to account. Businesses and corporations hold the fate of the world's plastic crisis in their hands. And government needs to take action against the biggest plastic polluters.

16 May 2022

The above information is reprinted with kind permission from Friends of the Earth.
© 2024 Friends of the Earth Limited

www.friendsoftheearth.uk

The myths surrounding single-use plastic

A ban on plastic cutlery, plates, and trays came into force in October 2023 – but are the 'sustainable' alternatives really that eco-friendly?

By Boudicca Fox-Leonard

From Sunday (1st October 2023), takeaways and restaurants in England are banned from handing out single-use plastic. New regulations mean that they can be fined £200 if they are caught giving customers plastic cutlery, polystyrene cups or containers.

And, although many are hailing it as another vital step towards the total elimination of plastic waste in Britain, there have been warnings that many businesses are not prepared and that the new rules will be too expensive to enforce.

But there is an even more fundamental question that needs to be asked: will the change actually benefit the environment?

Because the truth is, although it has been demonised by the green lobby and companies are under huge pressure to reduce or replace it, plastic is not necessarily the toxic material it is made out to be.

A decision by Lego this week encapsulates the problem. When the Danish toy company first announced its intention to make its bricks from recycled bottles, it was hailed as a huge step in sustainability. But the manufacturer has now said that it is abandoning the plans, saying that switching from oil-based plastic to recycled materials would have led to higher carbon emissions over the product's lifetime.

Lego's chief executive said that the company had tested 'hundreds and hundreds' of methods, but could not find the 'magic material' to solve sustainability issues.

Using recycled bottles required greater energy for processing and drying, and extra ingredients for durability.

The backtrack by the world's biggest toymaker highlights the extent to which plastic has been demonised, with companies under huge pressure to reduce or replace, even when it has little economic – or environmental – benefit.

James Piper, a recycling expert and author of *The Rubbish Book*, had predicted that Lego's initial idea was an environmentally friendly pipe dream that would do more harm than good.

'There is an element of companies knowing plastic is a good product, but that's a very tricky message to get across to the consumer,' he says.

Piper had foreseen another problem: Lego would have disrupted the existing market for recycling polyethylene terephthalate (PET), a polymer found in items such as bottles, packaging and some textiles, forcing up the price and making recycled PET plastic unaffordable for manufacturers of drinks bottles.

'A tonne of Lego is worth thousands of pounds, much more than plastic bottles,' explains Piper.

'You would suddenly have a massive distortion in the market, where all the drinks companies, Coca-Cola and Pepsi, are looking for recycled PET and can't get hold of it because Lego is buying it.'

The 'capture rate' of Lego – the proportion that is recycled – would also have been very low.

As a high-value product, the bricks are designed to last for many years, taking pride of place on your shelf at home, or – for many despairing parents – strewn all over the carpet.

Only a small percentage would likely have been recycled compared to the percentage of a single-use item such as a drinks bottle, which is recycled over and over again.

And Lego is not the only company to discover that recycling can have unintended consequences.

Marks & Spencer's move this month to replace its plastic 'bags for life' with paper carrier bags may sound environmentally friendly, but Piper says, it could actually have the effect of causing more plastic to go to landfill.

'Already most supermarkets manufacture their plastic bags from recycled shrink wrap used to deliver their products to stores,' he explains. 'If everyone moves to paper, that shrink wrap doesn't get recycled – there's no market.' (Not to mention the fact that paper bags are less well equipped to carry heavy loads of shopping, especially if they become wet.)

Plastic is such a toxic topic that, even when there's a good reason to use it, such as less wastage, lower cost, or relative ease of recycling, it's a hard sell.

'Instead of demonising it, we should be questioning the purpose it's being used for in the first place,' says Piper. 'Do we need [the product]? And if yes, what material should it be made from?'

Why a plastic milk bottle is greener than a carton

Plastic milk bottles are the poster child for recycling done well. 'As far as I'm aware, every council in the UK collects them, so that's a perfect story,' says Piper, of the HTP plastic bottles widely used in the UK.

However, this good news story gets overlooked – there's a perception that Tetra Pak cartons, which are plastic and paper welded together, are preferable. 'They are very difficult to recycle and very few councils collect them,' says Piper.

'However, it looks good because on the outside it looks like cardboard. So everyone thinks it ticks a sustainability box.'

Meanwhile, glass bottles need to be reused about 20 times to be sustainable. 'Some companies are reporting getting 30 uses, so that's pretty good,' says Piper. 'In a refill context, a glass bottle works. But if you are using it once and crushing it, then it's not.'

Plastic wrapping prevents food waste

If you've ever left a cucumber to perish in the bottom of your chiller drawer, you'll know the life-enhancing power of plastic to delay the sludging process.

In fact, plastic makes up about 5% of a product's sustainability credentials. 'While we all focus on the packaging, that's only a small part of the sustainability story,' says Piper. Food waste is a huge sustainability issue and plastic on cucumbers makes a difference.

In 2017 a study of data from retailers in Australia showed a food waste rate of 9.4% among cucumbers not wrapped in plastic, compared with 4.6% for those wrapped.

The trouble with paper straws

Stories about how they accumulated with other plastic waste and formed huge floating masses on the ocean surface drove a decision to ban plastic straws in 2020.

Since then we've been left to suck it up through soggy paper alternatives. 'There's little doubt that they are difficult to use and tend to disintegrate,' says Piper. While he doesn't think we should bring back plastic straws, he also stresses that banning them has led to paper straws seeming like an ethical decision, even when they're not.

'There have been lots of studies done that show that if something is made of paper people believe it's more environmentally friendly,' he says. This means consumers don't question using paper straws, even though they are wasteful too.

On top of this, some studies have shown that paper straws contain polyfluoroalkyl substances (PFAS), commonly known as 'forever chemicals'. This not only means that the straws likely aren't biodegradable – they are also vectors for chemicals considered hazardous to human and environmental health.

Wooden cutlery may not be an ethical choice

Though the woody taste of a disposable spoon or fork can strip the joy out of lunch, they seem like a more conscious choice.

However, as Piper says: 'There's never a recycling bin for wood when you're out on the go, so that's got to go in the bin.' At which point, is recyclable plastic cutlery so bad after all?

Compostable options might seem like a sensible solution, but current recycling systems aren't configured to deal with it, so more often than not it ends up in landfill anyway.

'We need restaurants and takeaway outlets to think through the end of life. When they use compostable cutlery there's no way of processing it.'

Again, Piper questions whether we should even have the option of the two – far better that we all start to carry our own metal cutlery and make a better choice for the planet.

The green credentials of coffee pods

Nothing seems more wasteful than aluminium coffee pods, but some research suggests they can be an environmental good, even when made from plastic.

While recyclable aluminium pods are more environmentally friendly than plastic, pods in general are better than filter or drip because they promote less coffee waste. Taking into account the whole life cycle of the product means greenhouse gas emissions, water and fertiliser are part of the sustainability story, not just the packaging.

For the record, freeze-dried coffee came out tops, environmentally, but is unlikely to keep coffee snobs content.

Why plastic can't be beaten for medical equipment

It's hard not to wince at the thought of how many tons of single-use personal protective equipment (PPE) the pandemic created. Little bits of plastic are the mainstay of the modern medical world.

'With medical equipment, the appeal is obviously that plastic is good for sterilisation,' says Piper. It highlights bigger questions about how plastic should be used.

'If you're going to have plastic anyway because it's a byproduct of the oil industry, let's have it in places where it doesn't disappear and get used once and never enjoyed or used again,' says Piper.

Where that's not possible, as with PPE, it's imperative to create recycling markets. 'That way shrink-wrap film becomes carrier bags. Or plastic bottles become plastic bottles. Let's not disrupt those markets.'

29 September 2023

The above information is reprinted with kind permission from *The Telegraph*.
© Telegraph Media Group Limited 2023

www.telegraph.co.uk

I avoided single-use plastic for a week – it should not be this hard or expensive

Greenpeace says that the UK produces more plastic waste per person than almost any other country. So why do shops still make it so tough for individuals to ditch the stuff?

By Emma Henderson

The cashier at the plastic-free shop is talking me through their incoming deliveries. 'Most come in paper or hessian sacks, but some still arrive with some plastic,' he says, a little meekly. Around me, the shop is packed with containers of dried food, confectionery and spices. They sell these items as refillable so customers can buy without plastic packets, but before they're on the shelf it seems that some things (like dried fruits) still arrive in them to keep them fresh.

I am not normally this interested in the logistics of packaging but I was in the middle of an experiment to try to live without single-use plastic – starting with just a week-long stint to test the waters. While at first it felt virtuous to be in this plastic-free shop, I was starting to worry that while the optics might be good, it might not stand up to real scrutiny.

The UK consumes a lot of plastic. In fact, Greenpeace says that the UK produces more plastic waste per person than almost any other country, with a truckload entering the ocean every minute, and UK supermarkets producing 800,000 tonnes per year. As well as turning the oceans into 'plastic soup', says Greenpeace, this plastic habit leaves a lot of our waste sent abroad. 'Most [waste] is going to countries that aren't equipped to handle it. Greenpeace investigators found British plastic waste being dumped and burnt in Turkey – on the roadside, near waterways and in the open air, and people nearby have reported serious health problems.'

Ideally, governments and large corporations would take huge steps to conquer our plastic problem but as individuals it also feels like we can help by not buying so much single-use plastic. But putting this sentiment into practice is certainly not that easy and neither is it affordable.

My first inroad into trying to live plastic free was in 2018 when I attempted Plastic Free July. At that time some companies were just starting to think about plastic free. My strongest memories of that time were giving up on finding a decent plastic-free shampoo, and struggling to stick to a refillable plastic-free deodorant (forgive me for not wanting to smell). It was undeniable that many things had a higher price tag too.

Four years on, as we continue to lurch from one climate-related crisis to another and the COVID-19 pandemic has caused a huge rise in the production of single-use plastic items like face masks and gloves, the need for plastic free feels more urgent than ever. But we are also facing a cost-of-living crisis: the price of basics is at its highest in over a decade, crippling families and squeezing their wages, meaning they are understandably prioritising spending less and spending carefully.

In this context, I wanted to know – has it been made any easier or more affordable for individuals to try to reduce plastic use? I set myself a challenge: for one week I couldn't buy anything new that contained plastic, but I could use plastic things I already had, which felt fair.

The first bad news was a plastic-free week meant no Terry's chocolate orange (not the foil but the plastic casing inside the box), no hash browns for Saturday brunch, and my favourite Friday night Indian takeaway was out of the question because it is delivered in plastic. As if this wasn't hard enough to swallow, I realised I was also unlikely to find my other true love – cheese – sans plastic unless I paid through the nose at an upmarket deli.

Ignoring those hurdles, I knew if this was going to be a success, I needed to stock up on snacks and store cupboard basics so I didn't end up relying on last-minute options and finding that there were none. My first stop was the plastic-free shop: in lieu of my Terry's I bought some chocolate orange buttons, but 200g cost me just shy of £7. Gulp. In true British form, I smiled through gritted teeth at the cashier as she scanned it. I could only think a slightly smaller (112g) 'share bag' of buttons would be on offer for £1 somewhere. I savoured these like gold.

Next stop – savoury snacks. Chilli rice cakes, to replace Snack A Jacks, were 1.8 times more expensive. The cheapest pasta was organic penne at £1.33 for 500g. In my normal Sainsbury's shop, I'd buy the 500g pack for 85p, but comparing like for like, the organic bag would be only 13p less. Basmati rice was £2.05 for 500g compared to £1.35 at Sainsbury's.

With cupboard essentials for the week, and some treats, I paid just under £16. But I would need more food.

My next stop was the small Saturday vegetable stall on the high street. I was prepped with veggie bags – the irony of them being made of nylon was not lost on me, but they're deemed better than single-use plastic vegetable bags, so onwards I went. The two trestle tables had metal bowls bulging with produce, everything from red peppers to aubergines – and even figs – and okra. The owners said it runs until everything is sold, so little of it is wasted. Tick, tick, tick.

But the raspberries and blueberries I had come to buy – clad in plastic punnets. No luck.

Then I wondered – where on earth does sell berries not in plastic punnets? (If you know, answers on a postcard, as this has left me totally stumped.) Determined for the market trip not to be a total failure, I grabbed aubergines for an aubergine parmigiana before clocking that the key ingredient – mozzarella – also comes in a plastic bag. Without cheese, it instantly loses its appeal. Another item not allowed this week, the list was already starting to mount up, just hours in.

During the working week, I found the biggest hurdle was office lunches. Working in central London, you'd think it would be easy to find a plastic-free lunch, but no. Nearby places like Pret and Wasabi's packaging are full of it and Leon serves most hot food in plastic-lined containers. Scanning the menu, I could have had waffle chips, but without a side of mayo (in a little plastic pot) they seemed pretty uninteresting.

Eating out wasn't all doom and gloom, though. One evening I went to a restaurant for dinner, and didn't catch sight of any plastic water bottles or an offer of straws, instead being given tap water in bottles. But I couldn't help thinking about the plastic that was very likely in the kitchen, from deliveries and other sources. Similarly to the shop, it was beyond my control.

One big mistake to confess: going to a bakery for a weekend treat, I picked out a baguette from the counter, not a shred of plastic in sight, and as I turned around to choose a loaf of plastic-free bread (£3.50 and it wasn't even sourdough), the baguette had been slid into a bag with a plastic window. The cardinal sin. But with the filling already smeared across the inside, meaning the bag couldn't be reused, it was pointless to try to rectify it. I was so near to completing the week without any big hiccups, yet felt cheated by this plastic imposition.

Admittedly, after only seven days, I was already looking forward to buying a whole host of things: blueberries, yoghurt, hash browns, bacon, 'nduja sausage, biscuits,

> If you want to start reducing your single-use plastic, these tips can help:
> 1. Plan, plan and plan. Food is the hardest area, so get used to thinking ahead. It doesn't have to be set in stone, but it will also help reduce food waste.
> 2. Find your nearest plastic-free shop. Even if you only buy a few store cupboard essentials here, it can make a difference. Plus, the staff will have more good ideas for you and in some shops, you can click and collect.
> 3. Make plastic containers your best friend. Use it in refill shops, transport food in it, and use it to store leftovers in the fridge and freezer.
> 4. Consider subscribing to a Who Gives A Crap toilet paper, where you can get 24 or 48 rolls. It is more expensive, but it uses waste office paper instead of cutting down virgin trees.
> 5. If you use cotton pads for removing make-up, swap to the reusable Face Halo pads. They can go in the washing machine and last for years.
> 6. Using a glass bottle milk delivery service in your local area means you won't have to buy it in plastic, and you'll never run out.

cheese and cereals. It's a long list of things that were either very hard or impossible to obtain, or alternatives that were so much more expensive I couldn't justify them.

Not only were some things impossible or financially unjustifiable, but by the end of the period I was struggling to find enjoyment in needing to go to three shops to buy my ingredients, when I normally could go to one. For people with busy lives – full-time jobs, commuting, children, caring commitments – requiring individuals to spend hours doing these jobs across multiple sites means it is an unappealing prospect before you even consider the money aspect (of both the items and the additional travelling back and forth).

Did I enjoy the challenge of the week? Yes, I made more considered purchases at independent shops, which not only helps to reduce waste but supports small and local businesses. Although Sainsbury's is convenient (it is only 20 metres from my flat), shopping there doesn't exactly nourish the soul, and the dominance of plastic, even on the fresh produce, highlights how completely detached we are from our food's roots.

But could I be plastic free full time? No. Not even close with the current provisions. Not being able to buy hash browns is unfortunately a bridge too far for me. Joking aside, the bottom line is that being plastic free in the UK remains largely inconvenient, and it can be prohibitively expensive. Combined with the cost-of-living crisis, it is a huge barrier, which is not only true for me only buying for two, but what about a family of four, five or more? If even the plastic-free shop can't wean themselves off entirely, how am I supposed to?

25 October 2022

The above information is reprinted with kind permission from *i* News.
© 2024 Associated Newspapers Limited

www.inews.co.uk

What are microplastics and why are they a problem?

From the deepest parts of our ocean to the summit of Mount Everest, microplastics have spread to nearly every crevice on the planet.

By Charlotte Cameron

Many are aware of the wider impacts of plastic pollution on our environment and there have been significant behavioural changes towards phasing out single-use plastics. The recent introduction of the Plastic Packaging Tax (April 2022) in the UK has offered greater financial incentives for businesses to move towards recycled plastics, while at an individual level we are more aware of recycling and are opting for plastic-free alternatives. Global campaigns have helped with this, illustrating the impacts that discarded plastics have on wildlife, from images of sea turtles ingesting plastic bags to birds caught up in plastic packaging.

But studies into the impacts of microplastics in humans and in food are only just emerging.

What we do know is that we come across and ingest microplastics every day – in our drinking water, in the meat, milk and blood of farm animals, in table salt, and in beer.

What are microplastics and where do they come from?

Microplastics are small pieces of plastic that are less than five millimetres in length, or the size of a small sesame seed.

They come from a variety of sources and are often broken down into two categories: primary and secondary microplastics.

Primary microplastics are the particles designed that way for commercial uses such as microbeads in facial scrubs or microfibres in textiles. Secondary microplastics are from the breakdown of larger plastics such as water bottles. The breakdown of this plastic is caused through environmental factors such as exposure to wind or sun.

The issue with this is that their breakdown doesn't just end in microplastics. Instead of breaking down into harmless molecules they decompose into particles similar to dust called nanoparticles.

Plastic pollution has invaded our oceans mainly due to littering but also because of storms, water runoff and winds that carry plastic of all sizes into our ocean. It is estimated that there are 24 trillion pieces of microplastics in our ocean which weighs the equivalent of roughly 30 billion 500ml plastic water bottles. Some research suggests that by 2050 there could be more plastic than fish in the sea by weight.

But microplastics are not turning up solely in our oceans. Now, studies show that they are found in human blood, our drinking water, and in the meat and milk of farm animals.

Why are microplastics bad?

All plastic pollution is detrimental to our environment; however, due to their size microplastics pose a particularly difficult challenge. Their size means they are more likely to get into our food chain and the result is that the chemicals end up in our food, our water, in our wildlife, and in us.

The health impact of microplastics on humans is still relatively unknown and hard to assess as each bit of microplastic contains a different combination of chemicals. Some research suggests that humans are consuming around a credit card's worth of plastic a week.

More studies have been done into the impacts of microplastics on our environment and wildlife. In July scientists at the Vrije Universiteit Amsterdam (VUA) in the Netherlands found microplastics in beef and pork for the first time, as well as in the blood of cows and pigs on farms.

They were found in every sample of animal pellet feed tested too, which is a possible cause for contamination. The food products itself were packaged in plastic, which scientists are suggesting is another possible route. Using the same methods to test the animal products, the VUA also reported on microplastics in human blood.

One thing is certain. The problems microplastics cause will only intensify as around 400 million tonnes of plastic is produced each year and that is expected to double by 2050. The challenge is that even if we eventually phase out most plastics – existing plastics will continue to degrade and at the moment this is a mass estimated at around five billion tonnes.

Tech that is removing microplastics from our environment

With advancing innovation and technology, new solutions are emerging to tackle the growing microplastics crisis.

Bacteria

Studies, such as one done by Dr Juan José Alava, are looking into bottom feeders already in our oceans, from sea cucumbers to tiny organisms, that could act as 'living vacuum cleaners'. These strains of bacteria are able to break down synthetic material. 'The idea is to identify communities of bacteria and try to enhance them – not by incorporating a new mix of genes created by humans, but by stimulating them to break down plastic,' Alava says.

Magnets

Fionn Ferreira, a 21-year-old Irish inventor, came up with a solution to magnetise the removal of microplastics. Winning the top prize at the 2019 Google Science Fair, Ferreira's homemade 'ferrofluid', which is a mixture of oil and powdered rust, successfully removed 88% of microplastics from water samples. In the future, this could act as a self-cleaning filtration device that could be used in ocean engines.

Plant-based nets

A new type of water filters made from a plant-based mesh is helping to filter out even the tiniest of plastic particles: nanoplastics. This technology can help researchers to study nanoplastics while also removing them from our oceans.

How to avoid microplastics

Although the human health impacts of microplastics are relatively unknown at this stage, many people are looking for ways to avoid consuming and producing microplastics in their everyday lives.

Here are just five ways:

1. Switch up your laundry routine
Many of our clothes are made from plastic particles. Polyester, nylon, and other synthetic fibres make up around 60% of the material that make up our clothes. When washed or dried, tiny particles can become dislodged. Opting for air drying where possible or washing settings of a lower speed can help stop this from happening.

2. Use plastic-free cosmetics
Personal care products, like face scrubs and toothpastes, are often made up of plastic components like microbeads. Look to see which beauty brands avoid plastic in their products.

3. Use public transport
Car tyres are a major source of microplastics as friction causes tyres to break down and shed particles. In busy cities with lots of traffic this leads to 'city dust', which is an accumulation of plastic particles and other debris. Using public transport can help avoid this tyre erosion.

4. Research clothing materials
You can avoid microplastics from your clothing by opting for garments made from natural fibres – cotton, linen or hemp, for example.

5. Advocate for single-use plastics
Most plastics will eventually break down into microplastics over time so one of the best ways to help tackle this crisis is to avoid single-use plastics altogether. Opt for a tote bag over a plastic shopping bag or a reusable cup instead of a throwaway takeaway cup for your coffee.

Companies tackling plastic pollution

While innovative solutions are emerging to tackle existing plastic pollution, an increasing number of companies are adopting plastic-free alternatives within their business to stop the root cause.

One of these is Flexi-Hex, a company providing an innovative packaging sleeve made from recycled paper that protects fragile products in transit. It acts as an alternative for unsustainable packaging like bubble wrap or air-filled plastic, helping to minimise excess waste and plastic.

The company was started by two surfers looking to solve a problem. How can they transport their boards safely while minimising their impact on the environment? Since then, the company has prevented over 3,166km of plastic packaging from being used, the equivalent to, as the crow flies, Cornwall to Turkey!

Companies are experiencing a real shift in consumer demand, with over two-thirds of UK consumers wanting to see greater use of paper-based packaging which can go into kerbside recycling collections, and 36% of Europeans boycotting brands over packaging sustainability concerns.

'Our sleeves have been designed with the consumer in mind, offering an elevated unpacking experience that is both planet-friendly and luxurious without compromising on functionality,' Flexi-Hex said.

The Beeswax Wrap Co is another company born out of a gap in the market for plastic-free packaging alternatives.

Their founder, Fran wanted a more sustainable alternative to cling film to use in the kitchen and so developed a wrap made from British beeswax and organic cotton. Now, The Beeswax Wrap Co offers a collection of sustainable plastic-free products for the home, all handcrafted in a solar-powered workshop in the Cotswolds.

Not only are these companies working to provide a sustainable solution to help tackle plastic pollution, they are also working to measure and reduce their own carbon emissions through Planet Mark certification with a reduction target of 5% annually.

We all play a role in ensuring we are not contributing to the plastic pollution crisis. This requires education and awareness, innovative solutions, and collective action across individuals, organisations, and governments.

The above information is reprinted with kind permission from PlanetMark.
© PlanetMark 2024

www.planetmark.com

'They're in the air, drinking water, dust, food...' How to reduce your exposure to microplastics

No corner of the planet is free from minuscule fragments of plastic packaging, textiles or utensils. We ask scientists what this means for our health – and what we should do to protect it.

By Amy Fleming

Invisible specks of eroded plastic from long-forgotten toothbrushes, sweet wrappers, and stocking-filler toys are everywhere. They live in our laundry bins, the Mariana trench and the human bloodstream. Microplastic particles can be small enough to infiltrate biological barriers such as the gut, skin, and placental tissue. We are all now partially plastic – but how worried should we be, and is there any way to minimise our exposure?

At the moment, says Stephanie Wright, an environmental toxicologist at Imperial College, London, a lack of epidemiological and in-human data means we don't yet know the harmful effects of microplastics, but 'I would say reducing particle exposure in general (including microplastic) is likely to be beneficial'. But avoiding the stuff is a tall order, considering it's in the 'air, drinking water, dust and food'.

Food and drink sealed in plastic has long been associated with cleanliness, purity and protection from contamination, but we now know that some of the highest exposures to microplastics, says Wright, 'are likely to come from processed and packaged foods and drinks'. The shedding of plastic is increased when containers are exposed to heat. 'Hot water in plastic-lined cups and takeaway containers also release micro- and nanoparticles, in some cases trillions per litre, although whether these are true plastic particles is unknown.'

Wright says that to reduce exposure to microplastics, 'I would start by not heating anything in plastic, or consuming hot liquid that has come into contact with plastic'. This includes microwaving food in Tupperware, or ready-to-heat products such as boil-in-the-bag rice and 'food-grade nylon used for food packaging, as liners for baking pans in restaurants and commercial kitchens, and in slow cookers in household kitchens'.

When it comes to water, she chooses tap over bottled: 'Some bottled waters – including glass bottles – contain thousands of microplastic particles per litre.' And, ideally, she would take it filtered. When I mention filtering to Mark Taylor, Chief Environmental Scientist at the Environmental Protection Authority in the Australian state of Victoria, he points out that home water filters are usually plastic, too: 'Ultimately it will start to shed because it will degrade.'

This gives me the perfect opportunity to gloat about my glass and stainless steel filter jug, but then I remember that the charcoal refills come in plastic pouches. When you start observing your plastic use, it's hard not to spiral. 'I think we can stress ourselves out over all of these things and put too much focus on it,' says Taylor. 'The reality is people are living longer than they've ever lived before. Some people in a [global] population of eight billion, of course, will be affected and may well die as a result of microplastics exposure.' The way forward, he says, is 'balancing the risk of microplastics versus practical actions and lifespan'.

Having extensively studied microplastic exposure in homes – which is where Taylor says we absorb the most plastic contamination – he knows it's impossible to avoid the stuff and so there's no point worrying over every bit of plastic we meet. Instead, he says, 'we can look at minimising inessential uses'.

At one end of the scale sits a plastic heart valve, which is essential. Whereas fruit sealed in plastic is unnecessary. 'You can think about the furnishings and the clothes that you acquire, and buy more natural fabrics,' says Taylor. 'Instead of having a polyester carpet, you could have a wool carpet.' Natural fibres are often more expensive, but second-hand is always an option, and if it's not something you can change, don't sweat it. 'You can think about buying natural clothing – they do produce microfibres, but they're not microplastics and they break down. If you've got kids, do you need to have plastic spoons and plates?'

On a personal level, he says, he makes choices based on unnecessary exposure, but also as an act of consumer protest – 'every little action matters'. It is often hard to find out the composition of plastic products – they don't come with ingredients lists like food – but he recalls looking for a new watch strap and discovering one contained perfluoroalkyl and polyfluoroalkyl substances (PFAS). 'I went: well, I won't be buying that. It is associated with testicular and kidney cancer, low birth weight in newborns and just a tonne of things.' PFAS are among the many commonly used chemicals in plastics that are endocrine disruptors, which some scientists believe are to blame for declining global sperm counts.

He actively avoids buying food such as fruit and veg wrapped in plastic, or adorned with 'those stupid little food stickers'. His household uses glass instead of plastic in the kitchen.

'I wear predominantly, but not entirely, natural fibres, because my work jacket is made of polyester. But I prefer either cotton or wool.' He concedes, however, 'I have a wooden floor with varnish, which I know will slough off.'

Keeping a clean house is something anyone can do to reduce exposure. 'The carpets, the curtains, the sofa, most of those are probably not made from fully natural fabrics, and they degrade and their fibres accumulate,' he says. All that dust and fluff that balls up like tumbleweed under sofas, or twinkles in sunbeams after you plump a cushion, will contain plastic fibres. This is why the vacuum cleaner is about more than being house-proud.

He says: 'It's very clear, whether you're dealing with microplastics or trace metals such as lead, zinc, cadmium and arsenic that migrate into a home, that regular vacuuming is really effective at reducing the load.' If you don't vacuum, the dust remobilises and, adds Taylor, 'deposits in open water vessels, on your fruit, on people's hands, kitchen utensils'.

He recommends – if you can afford it – robot vacuum cleaners, 'that go around the floor and just keep on top of the worst of it when you're out at work. Or preferably, if you've got a hard floor, wet mopping.' With carpets, vacuuming has the added benefit of capturing loose fibres soon to be shed from everyday wear and tear.

Malcolm Hudson, an associate professor in environmental science at the University of Southampton, is very keen that we don't panic about our current exposure to microplastics.

Instead, he'd rather we divert that energy into helping to stop the planet accumulating yet more plastic. At the current rate of production, more than 10 billion tonnes of mismanaged plastic waste will be dispersed in the natural environment by 2050.

He certainly isn't panicking right now. 'I'm sitting at home in my office and I'm probably breathing in some plastic fibres from the clothes that I'm wearing, and from the carpet on the stairs just outside my office,' he says. 'And I've probably ingested some plastic in my lunch, which is an unsettling thought but it's probably not doing me a great deal of harm,' he says.

He doubts, at this point, whether trying to limit plastic exposure will make much difference to his health right now.

'We've evolved to deal with inhalation and ingestion of impurities,' he says. 'That's why we have complex respiratory systems and all sorts of trapping devices to stop particles going into our lungs. It's why we have an immune system that's set up to deal with small foreign bodies. It's why we have a digestive system that doesn't let larger impurities get into our system – they just pass through.'

But in another few decades, 'if the environment continues to get more contaminated, I think you have got potentially a harmful issue.' This is partly due to the sheer volume of microplastics that will have accumulated by then, and we know that the greater the exposure, the greater the risk. 'There was a study from a few years ago that showed that people who work in textile factories in Bangladesh have been exposed to very high levels of airborne microplastic fibres and they do get respiratory disease.'

The other reason the health risks will grow with time is because the older the particles are, the more toxic they can become. They can harbour pathogenic microbes and take on other pollutants such as heavy metals. 'And then,' says Hudson, 'if you swallow that microplastic, you're swallowing a small dose of another harmful chemical as well.' These chemicals include, 'polyaromatic hydrocarbons, plasticisers like phenol A that are used in things like furnishings and packaging – they can have hormone mimicking or carcinogenic properties. Heavy metals like copper, vanadium, mercury, lead. Cadmium contaminated sediments have already become associated with plastics.'

Meanwhile, avoiding traffic-heavy roads is always recommended, where microplastics are part of the toxic soup of pollution, although Hudson reckons they're probably the least of your worries next to car fumes and tyre particles. Plastic comes off road markings and wears off brakes, says Hudson, 'made from composite synthetic polymers'. Roads are, adds Wright, 'a hypothesised source of microplastic particle emissions to the air due to litter being worn down and run over'.

But it's hard and time-consuming to prove the effects of any one pollutant on health. 'In a study, isolating the impact of microplastics versus all the other contaminants such as air pollution would be really difficult,' says Taylor. But rather than sit back and say there's no hard evidence they cause harm to humans, he says he would rather, 'apply the precautionary principle: in the history of environmental toxicology, early concerns were usually borne out. So let's take an approach that minimises – I don't think we can eliminate – the risk.'

10 July 2023

The above information is reprinted with kind permission from *The Guardian*.
© 2024 Guardian News and Media Limited

www.theguardian.com

Plastic and the Environment

Chapter 2

Myths about plastic pollution are leading to public confusion: here's why

An article from The Conversation.

By Lesley Henderson, Chair professor, University of Strathclyde

Does the prediction that there could be 'more plastic than fish in the ocean by 2050' concern you? How about reports that 'we eat a credit card's worth of plastic per week'? These are some of the 'facts' about plastic that are cited by the media.

They are certainly compelling sound bites and help to focus public and policy attention on the pressing topic of plastic pollution, but their scientific basis is far from robust.

The scientists whose findings were used to support the 'more plastic than fish' claim refuted this. But one scientist who worked on the original source the estimation is based on has now updated his figures. The claim is further undermined by the assumptions the calculation is based on an underestimate of fish stocks.

Research has also found that humans ingest less than a grain of salt of microplastics each week. This means that it would take around 4,700 years to ingest an amount of plastic equivalent to the weight of a credit card.

Over the past three years I've been interviewing households in the UK, Spain and Germany about plastics as part of a project focused on improving the recycling of plastic packaging.

I've been struck by the level of confusion people have about the sources of and risks associated with plastic pollution.

So, in collaboration with the Hereon Institute of Coastal Environmental Chemistry and communications experts, I have launched an online resource called 'Plastic Mythbusters' that aims to debunk popular plastic myths that regularly feature in media.

Negotiations are currently under way in Nairobi, Kenya, at the UN Environment Programme headquarters, to develop a legally binding global plastics treaty that covers the full life cycle of plastics – including their production, design and disposal. The Scientists' Coalition for an Effective Plastics Treaty – a network of independent scientific and technical experts – are calling for decisions to be based on robust evidence.

The focus of the negotiations is understandably on research from the natural sciences. But what role does media play in shaping public and policy responses to the plastics crisis?

Images of plastic pollution

The images of plastic pollution that are sometimes used by media are emotive and powerful, reaching vast numbers of people. The BBC's *Blue Planet II*, which was broadcast worldwide in 2017, showed audiences the impact of plastic waste on the oceans through upsetting scenes. One scene depicted a pilot whale carrying her dead newborn calf, which narrator Sir David Attenborough said possibly died because the mother's milk had been poisoned with plastic.

Scenes such as these are now synonymous with plastic pollution. They can raise awareness of the problem and help to shape the discourse on environmental policy.

After *Blue Plant II* aired, online searches for 'dangers of plastic in the ocean' increased by 100%. Michael Gove, UK environment secretary at the time, said he was 'haunted' by images of the damage done to the world's oceans shown in the series and then introduced a series of proposals aimed at cutting plastic pollution.

However, there is no clear evidence in the *Blue Planet II* sequence that the mother's milk was actually contaminated with plastics. Imagery such as this can also promote the idea that plastic pollution is a problem far removed from our everyday lives and that our actions, whether it be dropping plastic litter or engaging in local clean up initiatives, will have little effect. There is still no robust evidence linking *Blue Planet II* to a sustained change in people's behaviours.

Sidelining issues

The way in which the media presents the issue of plastic pollution can shape the preference for certain solutions and sidelines others. For instance, many people believe that the Great Pacific Garbage Patch – a large collection of marine debris in the North Pacific Ocean – is a solid mass. Framing the problem in this way assumes that plastic pollution can be removed from the ocean with the correct technology.

However, scientists describe the Great Pacific Garbage Patch as more akin to a 'growing plastic smog' that does contain larger plastic items but is also composed of trillions of micro and nanoplastics spread over large distances.

Experts point out that technical fixes are not always the answer, particularly where plastic is spread over huge areas resembling a very thin 'plastic soup'. In such cases, technical fixes are less practical, especially when considering the continuous addition of more plastic due to unchecked production.

Power of media to set the agenda

There is a growing consensus advocating for investment in measures to curb plastic production, rather than investing in costly technical clean-up efforts. However, by emphasising the individual responsibility of consumers to, for example, avoid single-use plastics, media coverage can divert conversations away from reducing plastic production.

The connection between plastics and climate change, or the impact of plastics on global biodiversity loss, are also not often covered by the media as much as emotionally charged images depicting marine animals entangled in plastics.

The original focus of the global plastics treaty was on marine litter, but it now encompasses the full life cycle of plastic pollution on all ecosystems. This includes plastic pollution in the atmosphere, and in marine, terrestrial and high altitude environments. This wider scope opens up the opportunity to explore public perceptions of the full life cycle of plastics.

The media is an invaluable resource that can play a key role in shaping how people perceive various issues. However, while it can effectively highlight the dangers of plastic pollution, there is a risk that an excessive reliance on emotive imagery may distract away from the policy that is actually needed.

In response to this article, a BBC spokesperson said that there is significant scientific evidence that contaminants found in some plastics can accumulate in fish and be ingested by adult whales. Those contaminants are then passed on to the offspring through the mother's milk.

20 November 2023

THE CONVERSATION

The above information is reprinted with kind permission from The Conversation.
© 2010-2024, The Conversation Trust (UK) Limited

www.theconversation.com

'Deadly' marine plastic hotspots revealed

Scientists are calling for urgent global action to address the escalating issue of marine plastic pollution, as a recent study identified deadly ocean hotspots.

By Yasmin Dahnoun

Scientists have revealed areas of the ocean where some of the world's most threatened seabirds face the highest risk from potentially deadly plastics.

Seabirds often mistake small plastic fragments floating on the surface of the water for food, or ingest plastic that has already been eaten by their prey. Ingesting small fragments can lead to poisoning, internal injuries, and starvation for seabirds.

The 26-year study assessed the movements of 7,137 individual birds from 77 species of petrel, including Critically Endangered species such as the Balearic shearwater. Petrels are an understudied but vulnerable group of marine birds that play a key role in oceanic food webs.

The breadth of their distribution across the whole ocean makes them important 'sentinel species' – an early warning of broad threats to species and ecosystems when assessing the risks of plastic pollution within the marine environment.

Boundaries

The research team – led by BirdLife International in partnership with Fauna and Flora, the British Antarctic Survey, University of Cambridge, and the 5 Gyres Institute –found that the areas of the highest plastic exposure risk are in the Mediterranean and Black Seas, the north-east Pacific, north-west Pacific, South Atlantic, and the south-west Indian oceans.

Although the team admits that data was limited for some regions such as coastal east and south-east Asia, the Mediterranean and Black Seas together account for more than half of plastic exposure risk.

The scientists overlaid global positioning data taken from tracking devices attached to the birds onto maps of marine plastic distribution to identify the locations where birds are most likely to come into contact with plastics while foraging or migrating.

Species of petrel included in the study were given an 'exposure risk score' to indicate their risk of encountering plastic during their time at sea. A number of already threatened species scored highly, including the Critically Endangered Balearic shearwater, which breeds in the Mediterranean, and Newell's Shearwater, which is endemic to Hawaii.

It is the first time that tracking data for so many species have been overlapped with plastics distribution maps on a global scale. The results showcase the extent to which the impact of plastic pollution on marine species transcends national boundaries, highlighting the need for international cooperation to tackle plastic pollution.

International

Catrin Norris, Programme Officer For Marine Plastics at Fauna and Flora, said: 'Once plastic enters the ocean, it is immediately out of our control, and – just like petrels and other marine species – it will pass freely from one country's exclusive economic zone to another.

'Many petrels and other marine species are already in a precarious situation and continued exposure to dangerous plastics adds to these pressures.'

'Plastic floating in one country's territorial waters may have entered the ocean many thousands of miles away. Considering that a quarter of all plastic that marine birds are exposed to is found in international waters, it is clear that this problem belongs not to any one country, but to all. We need international collaboration to develop effective plastic policies that truly put an end to plastic pollution.'

Jo Royle, Founder and CEO of Common Seas said: 'Most humans will never see the plastic gyres floating in the ocean, but our wildlife does. As we continue to learn of the detrimental impacts of plastic, we must picture ourselves surrounded by mountains of plastic in our own homes and visualise what life must be like for wildlife confused by this foreign substance in their habitats.'

Plasticosis

Another recent study has identified a new disease caused by plastic that is having a devastating impact on seabirds: plasticosis. Scientists have defined this as a combination of impacts from ingested plastic. For example, when a bird's stomach is damaged, it is then unlikely to function correctly as a digestive organ or barrier to virus and bacteria. Other organs, such as the liver, can also be seriously harmed by exposure to nanoplastics, which causes chronic inflammation.

Dr Jennifer Lavers, who contributed to both studies, said that a detailed knowledge of when and where marine animals are most at risk of encountering plastics while at sea could not be more timely. 'We lack a comprehensive understanding of plastics' impact on species and ecosystems, but what we do know is deeply concerning.'

Pressures

Bethany Clark, Seabird Science Officer at BirdLife International, said: 'Many of the birds included in our study are already affected by a wide range of threats including climate change, being caught in fishing gear, competition with fisheries and invasive species.'

'While the population-level effects of plastic exposure are not yet known for most species, many petrels and other marine species are already in a precarious situation and continued exposure to dangerous plastics adds to these pressures. Coordinated action to stem the flow of plastic pollution is needed now to protect seabirds and other marine life around the world.'

The study is published in *Nature Communications* and was supported by more than 200 researchers working in 27 countries. It was an output of the Cambridge Conservation Initiative's Collaborative Fund for Conservation sponsored by the Prince Albert II of Monaco Foundation.

3 July 2023

Design

Design a poster warning of the dangers to marine life from plastics.

Consider...

What could you do to prevent plastic waste from entering the ocean?

Brainstorm

In small groups, brainstorm some ideas on what can be done with plastic waste in the ocean.

Consider what products could be made with the plastic waste if it can not be easily recycled.

How can your product help reduce or prevent further waste?

The above information is reprinted with kind permission from Ecologist.
© 2024 The Ecologist

www.theecologist.org

Guide to the effects of plastic pollution in the ocean

It is hard to miss the news reports on television and in the press showing the effects of plastic waste and the amount of plastic debris in the ocean. From single-use plastic bags to fishing nets there is a vast quantity of plastics that end up in the sea. Most people understand that introducing plastic waste into the oceanic ecosystem is harmful. However, not everyone is aware of the devastating impact the plastic problem has already had and is still currently having on both the marine environment and for human health.

There are many issues relating to plastic waste in the ocean and why it is a problem that needs to be tackled globally. In this article we will explore the issues and look at the effects of plastic pollution in the ocean and marine environment.

The effects of plastic pollution in the ocean

When plastic enters the ocean, gyres (a system of rotating ocean currents, tides, and winds) can transport that piece of plastic all over the globe. For example, plastic waste that enters the ocean in Europe can end up on a beach in Asia which is why all countries need to be tackling this issue together since it goes beyond just a local problem.

Fish, sea mammals, and sea reptiles such as turtles are also impacted by plastic waste in the ocean. For marine life, entanglement and ingestion are serious issues that can lead to death. The knock-on effect is that longer term this can impact the overall viability and health of the entire marine ecosystem.

One of the major problems with ocean plastic is that it breaks down slowly. Some pieces of plastic can take hundreds of years to decompose. As plastic breaks down it often turns into smaller particles of plastic, called microplastics, that are even harder to clean up and deal with.

The effects of microplastics in the ocean

What happens to the plastic items when they become marine pollution?

As mentioned, one of the key issues with plastic marine debris is that over time it can degrade into smaller pieces called microplastics. Microplastics are defined as being pieces of plastic under 5mm in size.

These microplastics and microfibres can enter the food chain and also wash up on our beaches and are extremely hard to clean up. Although organisations have been working on ways to filter out this kind of plastic, it is challenging and costly to do so.

Another form of microplastic are microbeads which are small plastic pieces that are often designed for cleaning or cleansing in beauty products. These small plastic beads are made from plastic types such as polyethylene, polypropylene, and polystyrene. Many corporates have removed these from their products having understood the problem that they cause (and due to legislations by governments).

In a number of countries including the UK, US, Canada, NZ and Australia, microbeads have been banned from sale, but as highlighted, plastic contamination in the ocean is a global problem which requires a combined, global effort to address. Sadly, the damage has already been done and these plastic particles have already been introduced into the oceans.

The problem with plastic pollution on beaches

Anyone who has walked along a beach that has not been litter picked will often notice the quantity of rubbish and waste that is washed up on the shore. Often this waste will consist of flotsam and jetsam from commercial enterprises out at sea such as fisheries, but often you will also find bottles and packaging from household waste.

Unsightly debris on the shore visibly highlights the problem with poor waste management, but it is also only one of the many issues that rubbish has on the overall environment.

Seabirds, sea turtles, seals and other animals that occupy beaches and coastal areas can be harmed by plastic rubbish. From animals being entangled in fishing gear and nets, to ingesting plastic after mistaking it for food, are only two of the issues that plastic pollution in the ocean causes.

The problem with plastic on the ocean surface

Just as with plastic found on beaches, plastic that is floating on the sea surface can also be problematic. Marine mammals can get tangled in fishing nets and other marine species mistake plastic pieces for food and ingest it. Plastic ingestion can lead to suffocation, starvation, drowning and death.

Plastic on the ocean surface can also contribute to climate change. The reason for this is that with concentrations of plastic on the sea surface, bacteria that live in the oceans can attach themselves to the microplastics. The more plastic, the more bacteria. Since the bacteria consume more oxygen, they also in turn generate more carbon dioxide, which directly contributes to climate change since it upsets the biogeochemical carbon cycle.

The issue of garbage on the ocean surface can be visually dramatic. For example, The Great Pacific Garbage Patch, located between Hawaii and California in the North Pacific Ocean is estimated to be as large as 1.6 million square kilometres in size. While the ocean clean-up has begun in this area, the size and location pose a difficult challenge. However, merely cleaning up the ocean surface won't solve the issue of plastic in the ocean. This is because surface plastic is only the beginning when it comes to the plastic problem.

The impacts of plastic in the ocean

A less obvious and visual issue with plastic in the ocean is the plastic waste that either floats within the layers of the sea itself or rests on the seabed.

It is estimated that there is 14 million tonnes of small plastic under 5mm in size sitting on the sea bed. It is also estimated that the amount of plastic under the ocean's surface is 30

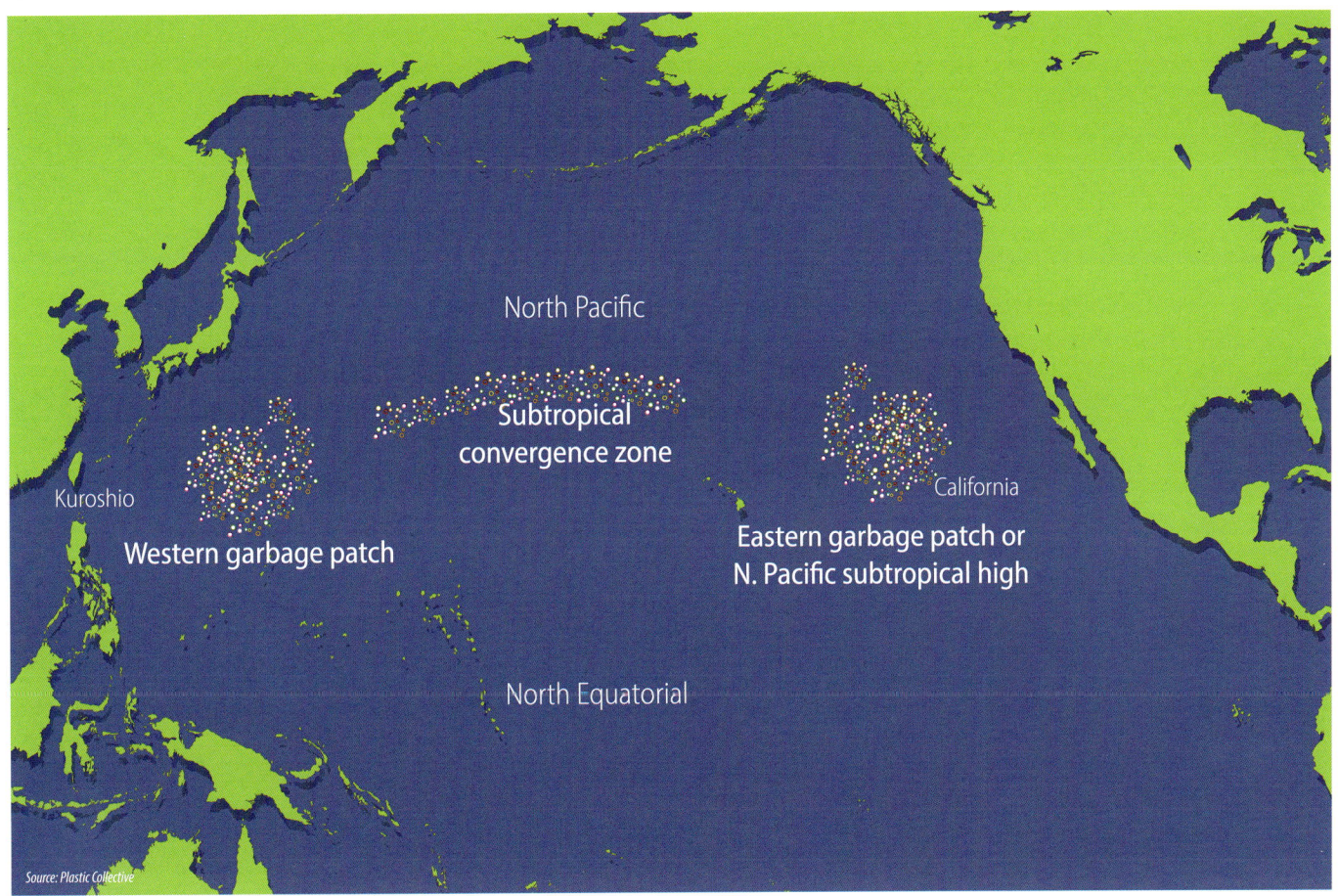

times greater than what is found floating on the top of the sea.

While marine mammals may die from the effects of ingesting plastic, smaller creatures such as fish and crustaceans can also eat plastic (often microplastics). This is also the case for plankton, a key food source for animals, which resides in the sunlit zone of the ocean.

When mammals and fish predate on smaller animals, if that animal has ingested plastic waste, the plastic is introduced into the larger animal's digestive organs. Plastic can have long-term health issues for animals. In molluscs there has been evidence of hormonal changes when exposed to plastic.

The wider implication is that since humans also consume fish, crustaceans and shellfish, it means that plastic, chemicals and microplastics are also in the human food chain.

Often these plastics are so small they go undetected and potentially can end even end up in the bloodstream. In a study carried out by a team at the Vrije Universiteit Amsterdam in the Netherlands, polymer particles have been found in 80% of the humans tested.

While the long-term effects of these plastic contaminants are unknown, it is theorised that these plastics could potentially cause cancer and other serious illnesses in humans. The bottom line is that plastic pollution is not only a significant problem for marine life, but it also has the potential to have a negative impact on human health.

Plastic can also have a negative impact on coral reefs, atolls, and biodiversity. It has been found that plastic contamination in coral reefs increases disease 20-fold.

What causes plastic pollution in the ocean

One question is where does all the plastic and rubbish in the ocean come from in the first place?

The majority of the plastic waste found in the ocean is from land-based household and corporate waste sources. This waste often enters the ocean from the runoff of rivers and waterways.

Often, plastic rubbish from single-use plastics, such as plastic bags, plastic bottles, straws, and other food packaging that has not been disposed of properly through recycling or in landfills can eventually find themselves in the waterways.

A smaller proportion of the marine plastic waste (but still a significant quantity) is from the fishing industry. Fishing nets and other fishing equipment can end up as ocean pollution and can cause damage to marine animals and habitats.

Of all marine pollution, plastic makes up around 80% of it and this equates to eight to 16 million metric tonnes of rubbish that enter the ocean each year. To put that in perspective it is estimated that there are between 50 and 75 trillion pieces of plastic in the ocean. This figure is set to increase, without significant intervention by consumers, businesses, and governments around the world to address the issue.

Long-term future

Without significant changes in the way we use and dispose of plastic the long-term future for the environment and oceans is bleak. It is predicted that plastic production will double by 2040 and increase by 2.5 times by 2050.

We are already seeing the effects on biodiversity, ocean habitats and the effects of plastic in the food chain for humans. Without significant intervention in our use and how we deal with the problem of plastic, the situation will only become worse. Without a fundamental shift, plastic waste in the ocean could nearly triple in quantity.

Currently only 20% of the plastic used is estimated to be recycled, leaving 80% of all plastic to have the potential for becoming a pollutant.

Back in 2017 the United Nations Environment Programme launched the Clean Seas Campaign to reduce problematic and avoidable plastics. Sixty-three countries pledged to support this improvement in plastics management, and while inroads have been made, much more needs to be done.

For the individual concerned about plastic waste and the environment, there are a number of minor changes that they can make to support the reduction in plastic pollution in the environment. It may feel like the impact at an individual level is minuscule but if everyone made the same changes it would have a substantial impact.

Individuals can support the reduction in plastic waste by firstly avoiding buying products that use single-use plastics, such as plastic water bottles. Another important thing is to ensure that the plastic waste you do produce is properly recycled. Helping to stop plastic from entering the environment is a key aspect of improving the problem with plastic pollution.

For corporates looking to improve their green credentials there are a number of things that can be implemented, and they are not always expensive or incur extra costs. Initiatives include the following:

- **Waste reduction** – Businesses should explore the waste they produce and look at ways to recycle and reduce the materials that they need to operate their business. Even small improvements to waste reduction totals can have an enormous impact over a year.
- **Use sustainable materials** – If you are a business that either manufacturers or ships products then switching to sustainable materials in your packaging and products can have a massive impact on how much extra plastic product waste is generated.

How Plastic Collective helps

Plastic Collective works with communities to tackle plastic waste. We supply education programmes and provide machinery and training to create sustainable plastic recycling micro-enterprises. We also provide a marketplace for communities to sell the recycled plastic they produce.

13 October 2022

Key Facts

- It is estimated that there is 14 million tonnes of small plastic under 5mm in size sitting on the sea bed.
- It is estimated that the amount of plastic under the ocean's surface is 30 times greater than what is found floating on the top of the sea.
- Of all marine pollution, plastic makes up around 80%.
- It is estimated that there are between 50 and 75 trillion pieces of plastic in the ocean.
- It is predicted that plastic production will double by 2040 and increase by 2.5 times by 2050.
- Only 20% of the plastic used is estimated to be recycled.

Brainstorm

In small groups, brainstorm some ways that you can reduce your plastic usage.

Design

Design a poster displaying how people can reduce their plastic usage.

The above information is reprinted with kind permission from Plastic Collective.
© 2018-2024 - The Pacific Collective Pty. Ltd.

www.plasticcollective.co

Pandemic face masks could harm wildlife for years to come

By James Ashworth

The protective equipment that kept us safe from COVID-19 could pose a lethal threat to nature.

While no longer as prevalent as they once were, the billions of face masks and gloves that were produced during the pandemic are making plastic pollution an ever-greater issue.

The impacts of COVID-19 go far beyond the disease itself, and are set to stay with us for centuries to come.

A study using community science observations from around the world found that disposable face masks and plastic gloves could pose an ongoing risk to wildlife for tens if not hundreds of years. Entanglements were one of the most prevalent threats, with some animals being killed after becoming caught in the plastic debris.

Dr Alex Bond, the Principal Curator and Curator in Charge of Birds at the Museum, was a co-author on the paper.

'Ultimately, we really don't know how big a problem pandemic waste could be,' Alex says. 'As many areas of the world had restrictions on non-essential movement, we will never be able to know the true extent of the issue, but this study gives us a snapshot into the sheer diversity of species that were affected.'

While the study captures only 114 observations from around the world, it is likely that it represents just a fraction of the much larger impacts of COVID-19 waste on wildlife.

With an estimated global demand of over 129 billion masks per month at the height of the pandemic, the effect of pandemic waste will become more pronounced as even more plastic works its way into our ecosystems.

'We filter out most litter in our environment, as it represents examples such as crisp packets or cigarette butts that we've seen for years or decades,' Alex adds. 'When PPE [personal protective equipment] flooded our waste management systems in the early days of the pandemic, it was a lot more obvious because it was new.'

'Now we don't even flinch when we see a blue face mask on the ground. It's rapidly become part of our everyday experience of waste in our environment.'

How did the COVID-19 pandemic affect plastic pollution?

When COVID-19 was declared a pandemic in March 2020, it kickstarted what the scientists describe as an 'unprecedented increase in the production and use of single-use plastics.'

The market value of the PPE industry jumped by around 200 times in the first year of the pandemic as legal requirements were introduced in countries around the world to stem the spread of the virus.

Some of these requirements specified particular types of face mask and other protective gear, most of which was single use. From March to October 2020, this caused the amount of abandoned face masks to increase by more than 80 times to represent almost 1% of all dumped litter globally, and as much as 5% in the UK.

Some of these masks even made their way to uninhabited areas, with 70 face masks found on the beaches of the Soko Islands believed to have washed ashore from nearby Hong Kong.

Disposable gloves, meanwhile, initially jumped to around 2.4% of dumped litter globally in April 2020 but then dropped back to 0.4% as the year progressed.

As levels of litter increased, wildlife struggling with pandemic-related debris became more common. For instance, the Royal Society for the Prevention of Cruelty to Animals (RSPCA) raised concerns after rescuing a gull which was having difficulty walking because of a mask tangled tightly around its legs.

The litter has also been linked to wildlife deaths, with one of the first believed to be an American robin found dead in Canada in April 2020 after becoming entangled in a face mask. Later that year, a face mask eaten by a Magellanic penguin in Brazil is believed to have led to the bird's death.

Even if not the direct cause of death, litter can weaken wildlife and make them more susceptible to fatal injuries. For instance, a gull in Rotterdam was struck by a car while tangled in a mask, which is believed to have limited its ability to escape.

Aside from birds, COVID-19 masks and gloves have also affected bats, crabs, hedgehogs and a variety of other wildlife.

The researchers behind the current study wanted to investigate the ways in which this debris has affected wildlife, and how community scientists could help investigate at a time when international travel was heavily restricted.

How was wildlife affected by COVID-19 waste?

The researchers used data from a variety of sources, including unpublished scientific reports, community science platforms and social media networks. The witnesses were then contacted to gain as much information about each pandemic waste incident as possible.

In total, the researchers gathered 114 sightings, of which 83% involved birds. Mammals were the next most exposed to COVID-19 waste, in 11% of sightings, while 3.5% of sightings featured invertebrates and 2% involved fish.

The results reflect existing research which shows that birds are at particular risk of entanglement from plastic, with an estimated third of seabird species and a tenth of freshwater species known to have been caught in synthetic items.

As a whole, entanglement reflects around 42% of the impact of pandemic waste on wildlife, but this is only slightly more than the 40% of sightings which saw masks and gloves used to build nests.

'Many birds build nests and they generally build them out of filamentous items, whether that's grass, twigs, moss or spider silk,' Alex says. 'Unfortunately, a lot of rubbish has the same characteristics, particularly objects like masks that have strings to loop around the ears. When that is incorporated into nests, it presents a significant entanglement risk to both adults and their chicks.'

Though the study was limited to English language searches, and social media platforms that aren't as dominant in some nations of the world, the research provides an insight into how the pandemic will continue to affect the environment.

With disposable face masks estimated to take up to 450 years to decompose, the waste left behind by the response to COVID-19 will have to be considered in any future attempts to tackle global plastic pollution.

In this fight against waste, the study demonstrated that community scientists could be relied upon as allies to help find and alert others to the problem.

The researchers called for greater efforts to develop more streamlined community science platforms to help people from all walks of life to assist scientists and policymakers in the battle against plastic pollution while making the field more equitable.

4 August 2022

The above information is reprinted with kind permission from The Natural History Museum.
© 2024 The Trustees of The Natural History Museum

www.nhm.ac.uk

New study into impact of plastic pollution on livestock and donkeys

Researchers will examine the amount of plastic cattle and donkeys ingest.

A new study by the University of Portsmouth and the Donkey Sanctuary in Lamu will examine the amount of plastic cattle and donkeys ingest, how this affects their health, and the impact on human lives. It will focus on animals on Lamu Island in Kenya, which is a UNESCO world heritage site where subsistence farming is common.

Researchers aim to gain a greater understanding of the severity of plastic pollution and find ways to solve it. They will work with animal owners and local communities in order to improve animal welfare.

Previous studies on the effects of plastic pollution on animals have focused on the marine environment rather than terrestrial animals.

Scientists already have an understanding of the effects that ingesting macro and microplastics can have on livestock. Ingesting macroplastics can lead to a loss of body condition and disease, and can also cause blockages in the digestive tract that can lead to colic, starvation and death.

Microplastics can leech into the soil and plants that the animals eat. This can lead to reduced growth and fertility, abnormal behaviour and biochemical and structural damage.

The ingestion of plastics can impact humans as well, with livestock often ending up in the food chain, and working animals such as donkeys, supporting the livelihoods of some of the poorest communities around the world.

Project lead, Dr Leanne Proops, said: 'In many countries, including Kenya, domestic animals graze open waste dumps to find food, consuming plastics that can have a dire effect on health and welfare. Even if livestock appear unharmed, meat and milk often contain microplastics that affect human health. The problem of plastic pollution is getting worse, and we need to know how this impacts the animals that play such a major role in communities in the Global South.'

Work has already begun to assess the foraging behaviours of cows and donkeys in Lamu. The research has identified that the ingestion of plastic by free-roaming domestic animals is a major problem, and initial results indicate that donkeys may be more susceptible to the effects of eating inappropriate materials.

Dr Proops continued: 'The next phase of the project will see livestock faeces analysed for plastic levels. This study will be the first to directly compare the relative risks of plastic ingestion in domestic species with differing foraging ecology and morphology. This is also the first step in quantifying the impact of plastic pollution on the welfare of livestock and equids in Lamu specifically, and will help to inform potential future waste management strategies.'

26 June 2023

The above information is reprinted with kind permission from MRCVSonline.
© 2024 Vision Media

www.mrcvs.co.uk

Littering: the effects on wildlife

By Jo Foster

Littering is deeply affecting our wildlife, with one billion seabirds worldwide are being killed by litter every year and 260 different marine species are trapped in plastic waste on a regular basis.

The RSPCA receives 5,000 phone calls every year regarding animals affected by litter – equating to an average of around 14 reports per day. The UK has the worst record for dropping litter in Europe. In England alone, a shocking 62% of residents have admitted to littering.

Globally, 1.5 million tonnes of plastic waste come from plastic water bottles, according to the World Wide Fund for Nature; and 33% of litter is caused by carelessly discarded fast-food wrappers. Cases of birds and animals being adversely affected by plastic litter have risen by 22% in the past four years alone, according to the RSPCA.

Trapped in plastic packaging

One of the biggest threats to wildlife is plastic packaging. Marine mammals are curious about plastic and, while investigating, they often become trapped. One of the most dangerous items is the plastic packaging that holds four cans of alcohol together. When carelessly discarded, it can harm many different creatures including turtles, birds, fish and small mammals. Large fish will often inspect plastic debris and be tempted to take a bite out of larger items.

Research in coastal regions has revealed 80% of plastic on the beach or in coastal waters shows birds' peck marks, suggesting they thought it was food. In particular, polystyrene foam is easily mistaken for cuttlefish bone that attracts birds. Unfortunately, ingesting plastic can prove fatal for fish, seabirds, and other wildlife.

Suffocating in plastic bags

Wildlife can suffocate in plastic bags and similar litter. Animals can confuse plastic bags for food or be tempted to look inside by a lingering smell of food, especially if it's been freshly dropped.

Small mammals can get trapped inside and suffocate or even choke if they try to eat the bag. If they swallow a plastic bag, it can obstruct their stomach or bowel, resulting in a painful and lingering death.

Cut by sharp objects

Wild creatures can cut themselves on sharp objects thrown away by humans. Empty cans are a source of fascination for animals including foxes and hedgehogs – they are likely to look inside for food.

Birds and marine creatures such as turtles can get tangled in fishing lines and hooks, resulting in serious injury and death. The RSPCA reports seals suffering from deep infected wounds, caused by sharp plastic objects cutting into their necks. There are also cases of geese and swans becoming trapped in fishing lines and netting.

Discarded chewing gum

You may not realise the impact that discarded chewing gum has on wildlife. However, if small mammals and birds tread on chewing gum, it can get stuck in their feet, fur, or feathers. It soon gets matted, making flight and movement difficult.

Damaged nests

Litter can damage the nests and homes that birds and animals have spent so long creating for themselves. Research reveals an increase in the number of birds mistakenly using plastic to build their nests – if they get tangled, there is a real danger they could die.

Seabirds taking food back to the nest for chicks and regurgitating it are inadvertently feeding their young harmful plastic mixed in with the food.

How to help

The easiest and best way to help is to simply stop dumping litter irresponsibly. Dispose of waste properly – there are further steps you can take, even if you recycle your waste.

Try to cut down on single-use plastics and reuse things before you throw them away. When it comes to specific types of waste, think carefully about how to make it wildlife-safe before you bin it. Tie a knot in the top of a plastic bag so no small animals can crawl inside and get trapped. Before disposing of the four-pack drink holders, cut each of the loops so they can't get wrapped around animals' necks. Similarly, cut elastic bands.

If you're an angler, fish responsibly and never leave nets or broken rods by the water. Take all waste home with you - leave no trace!

Crush metal cans before you send them for recycling and always wash containers and put the lid back on before disposing of them.

Benefits of recycling

The modern philosophy of 'reduce, reuse ,and recycle' should become a way of life if we are to save the planet's resources for future generations. Recycling saves energy and prevents excess greenhouse gas emissions and water pollutants.

Not only does it preserve virgin materials, but it also helps to reduce the pollution caused by extracting and processing. Refill bottles and reuse boxes; avoid purchasing anything packaged with single-use plastics where possible.

16 May 2022

The above information is reprinted with kind permission from Solent Plastics.
© 2024 Solent Plastics

www.solentplastics.co.uk

Solving the Plastic Problem

Chapter 3

15 ways to reduce your plastic use

Plastic is everywhere. It's in our phones, our clothes, our food packaging, our toys, and even our toothbrushes. Plastic is cheap, durable, and convenient, but it also has a huge environmental impact. Plastic pollution is harming wildlife, ecosystems, and human health. It takes hundreds of years to decompose, and it releases toxic chemicals into the soil and water.

According to the World Wildlife Fund (WWF), the world produces about 300 million tonnes of plastic waste every year, which is equivalent to the weight of the entire human population. Only 9% of this plastic waste is recycled, while the rest ends up in landfills, oceans, or incinerators. By 2050, there could be more plastic than fish in the ocean.

So, what can we do to reduce our plastic use and protect the planet? Here are 15 simple and effective ways that you can adopt in your daily life:

1. **Carry a reusable water bottle.** Instead of buying bottled water, which is often just tap water in a plastic container, bring your own reusable bottle and fill it up at a fountain or a tap. You'll save money, reduce waste, and stay hydrated.

2. **Say no to plastic straws.** Plastic straws were banned in England in October 2020, Northern Ireland in January 2022, Scotland in June 2022, and in Wales in October 2023. The UK had the highest plastic straw consumption in the EU, with around 8.5 billion plastic straws used each year. Plastic straws are one of the most common items found in beach clean-ups, and they can cause serious harm to marine animals that mistake them for food. If you need a straw, opt for a paper, metal, bamboo, or silicone one, or simply drink from the glass or cup.

3. **Bring your own shopping bag.** Plastic bags are another major source of plastic pollution, and they can take anything from 20 to 500 years to break down. Many countries have banned or taxed plastic bags, but you can also do your part by bringing your own reusable bags when you go shopping. You can use cloth, canvas, or jute bags, or even make your own from old clothes or fabrics.

4. **Avoid single-use packaging.** Wherever possible, choose products that have minimal or no packaging, or that use biodegradable or recyclable materials. For example, you can buy loose fruits and vegetables, or use your own containers to buy bulk items like grains, nuts, or spices. You can also avoid plastic wrappers, cups, plates, cutlery, and other disposable items, and use your own reusable ones instead.

5. **Recycle properly.** If you can't avoid using plastic, make sure you recycle it correctly. Check the labels and symbols on the plastic items, and follow the recycling guidelines of your local area. Some plastics, such as PET (polyethylene terephthalate) and HDPE (high-density polyethylene), are easier to recycle than others, such as PVC (polyvinyl chloride) and PS (polystyrene). You can also look for recycling centres or collection points that accept different types of plastics.

6. **Reuse or repurpose plastic items.** Before you throw away a plastic item, think of ways that you can reuse or repurpose it. You can use plastic bags to make rugs, baskets, or coasters. You can use plastic containers to store food, toys, or stationery. You can also donate or

sell your unwanted plastic items to someone who can use them.

7. **Choose natural or organic fabrics.** Many synthetic fabrics, such as polyester, nylon, and acrylic, are made from plastic fibres that shed microplastics when washed. These microplastics can end up in the waterways and oceans, where they can be ingested by fish and other aquatic animals. To avoid this, you can choose natural or organic fabrics, such as cotton, wool, silk, or hemp, that are biodegradable and eco-friendly.

8. **Switch to a bamboo toothbrush.** Did you know that every plastic toothbrush you've ever used is still somewhere on the planet? Plastic toothbrushes are not recyclable, and they can take up to 500 years to decompose. A better alternative is a bamboo toothbrush, which has a natural and compostable handle, and often has biodegradable bristles as well. Bamboo is also a fast-growing and renewable resource, and it has antibacterial properties that can keep your toothbrush clean.

9. **Use natural or homemade cosmetics.** Many cosmetics, such as face wash, body scrub, toothpaste, and lip balm, contain microbeads, which are tiny plastic particles that are added for exfoliation or colour. Microbeads are too small to be filtered by wastewater treatment plants, and they end up in the oceans, where they can harm marine life and enter the food chain. To avoid this, you can use natural or homemade cosmetics that use ingredients like sugar, salt, oatmeal, honey, coconut oil, or essential oils.

10. **Buy second-hand or eco-friendly products.** When you need to buy something new, consider buying second-hand or eco-friendly products that have less plastic or no plastic at all. For example, you can buy second-hand books, clothes, toys, or furniture, or you can swap or borrow them from your friends or family. You can also buy eco-friendly products, such as wooden or metal toys, glass or ceramic dishes, or paper or cardboard packaging.

11. **Make your own cleaning products.** Many cleaning products, such as detergents, bleach, and disinfectants, come in plastic bottles that are not always recyclable. They also contain harsh chemicals that can pollute the water and harm your health. To avoid this, you can make your own cleaning products using natural ingredients, such as vinegar, baking soda, lemon, or soap. You can also use reusable spray bottles, cloths, or brushes to apply them.

12. **Support environmental organisations.** There are many organisations that are working to reduce plastic pollution and protect the environment. You can support them by donating money, volunteering your time, or spreading awareness.

13. **Educate yourself and others.** One of the best ways to reduce your plastic use is to educate yourself and others about the problem and the solutions. You can read books, watch documentaries, listen to podcasts, or attend events that talk about plastic pollution and its impacts. You can also share your knowledge and tips with your friends, family, classmates, or social media followers. You can also join or start a campaign or a movement that advocates for reducing plastic use in your school, community, or country.

14. **Be a conscious consumer.** Every time you buy something, you are making a choice that affects the environment. You can be a conscious consumer by asking yourself some questions before you buy something, such as:

    - Do I really need this?
    - Is there a plastic-free or low-plastic alternative?
    - How long will I use this?
    - How will I dispose of this?
    - What is the environmental impact of this?

    By being a conscious consumer, you can reduce your plastic use and your ecological footprint, and also save money and resources.

15. **Start small and be consistent.** Reducing your plastic use may seem overwhelming or impossible, but it doesn't have to be. You can start small and be consistent, and you'll see the difference over time. You can start by choosing one or two ways to reduce your plastic use, and then gradually add more as you go along. You can also track your progress and celebrate your achievements. Remember, every little bit counts, and together we can make a big difference.

Write

Write a persuasive letter or email to a company, to encourage them to use less plastic in their packaging.

For example, you could write to a supermarket to ask them to use recyclable alternatives to plastic for their ready meals.

Brainstorm

In pairs, think about all the ways you can reduce plastic use at home. Write a list of all of your ideas.

Design

Design a poster with ways to reduce your plastic use as the theme.

Bioplastics: pros & cons and are they the future?

By Ben Hardman

Conventional plastic is one of the most successful products ever made. From an industrial perspective at least. Its use and versatility are unrivalled across the world.

But it has come with a heavy environmental cost.

Plastic takes hundreds of years to break down. Maybe even thousands. We don't truly know because every single piece of plastic that has ever been made still exists today in one form or another. As it does slowly break down, tiny fragments of plastic are released into the environment. These are called microplastics.

Microplastics have been found everywhere from the Antarctic to the world's deepest ocean trench and even in human blood for the first time ever.

It's clear that we need to move away from traditional plastic. One prospect stepping into the light is a bio-based plastic.

Could bioplastics be the answer the environment is looking for?

What are bioplastics and what are they made from?

Bioplastics represent a movement away from plastic made with fossil fuels such as crude oil and natural gas.

Instead, bioplastics are made from natural, renewable, and biodegradable sources.

It's possible to make plastic from natural materials as, in essence, plastic is a long chain of polymers capable of being moulded and shaped.

Bioplastics can be made from various plants and biological materials, including starchy vegetables and softwoods.

Some of the most common plants used for bioplastics include sugar cane, corn, wheat, and potatoes. It's also possible to synthesise bioplastics from microorganisms and algae.

Just like conventional plastic, the result is a versatile and useful material. In the modern world, there are plenty of uses for bioplastics, including kitchen utensils, containers, bags and bin liners, packaging, and even eco-friendly phone cases.

Pros of bioplastics

There are many positives that come with bio-based plastics, especially when compared to petroleum-based plastics.

1. Recyclable and compostable

One of the big issues with traditional plastic is that it can't always be recycled and will not break down into useful materials.

Bioplastics are biodegradable and will break down via natural processes. It's likely they'll have to be sent to an industrial composting facility where composting conditions – such as heat, oxygen, and moisture – can be carefully controlled to ultimately produce soil.

Creating recyclable and compostable materials means that waste will not be sent to landfills, which is not a long-term solution for the environment.

2. Made from renewable materials

If we are to create a more sustainable world, the use of renewable resources is going to be pivotal.

Renewable, plant-based resources offer a dual benefit to the environment: they are much less energy intensive than using crude oil, and they absorb carbon dioxide as they grow.

This means that renewable materials have a much gentler impact on the environment and greenhouse emissions.

3. Non-toxic substances

Being made from natural materials means that bioplastics don't contain unnecessary chemicals, toxins, and pollutants.

As bioplastics break down, they will not – or certainly shouldn't – contaminate the local environment. This makes them better for planet health, as well as that of marine and land ecosystems.

Cons of bioplastics

Despite all the positives, there are still some downsides to bioplastic.

1. Costly to create

As the bioplastic industry is new, the technology associated with it comes at a cost.

Specialist manufacturing processors and facilities are required to create bioplastics, and at the end of their life, industrial composters are needed to break the material back down into the soil.

It's thought that bioplastic costs two to three times more to create than traditional PET or PE plastic.

2. Lack of infrastructure to deal with bioplastics

Thanks to the cost of the equipment, many countries don't have the facilities yet to deal with bioplastics.

If material, including bioplastic, can't be recycled and processed properly, it usually means that it will end up in a landfill. If this is the case, then it's not too much better than a non-biodegradable type of plastic.

3. Land use issue

Creating tonnes of bioplastic requires many more times the land to grow the crops.

When lots of agricultural lands are needed, there are environmental and ethical questions that need to be asked.

Have previously forested areas been cleared for the crops, or should the land be used for food use instead? For example, should corn and potatoes being used for bioplastics be used to help solve hunger crises?

It's clear that bioplastics do offer a much brighter future than traditional plastic. They can help to lessen the impact on the environment as well as limit the release of greenhouse gases into the atmosphere.

However, some questions still remain about global adoption and whether we have the facilities to deal with bioplastics properly. If not, they'll just get treated like conventional plastic does.

6 April 2022

The above information is reprinted with kind permission from Green Living Blog.
© 2024 Greenlivingblog.org.uk

www.greenlivingblog.org.uk

£3.2 million for innovation in plastics reduction

Seventeen ground-breaking projects announced in March 2023 have the potential to alter the UK's relationship with, and management of, plastic packaging.

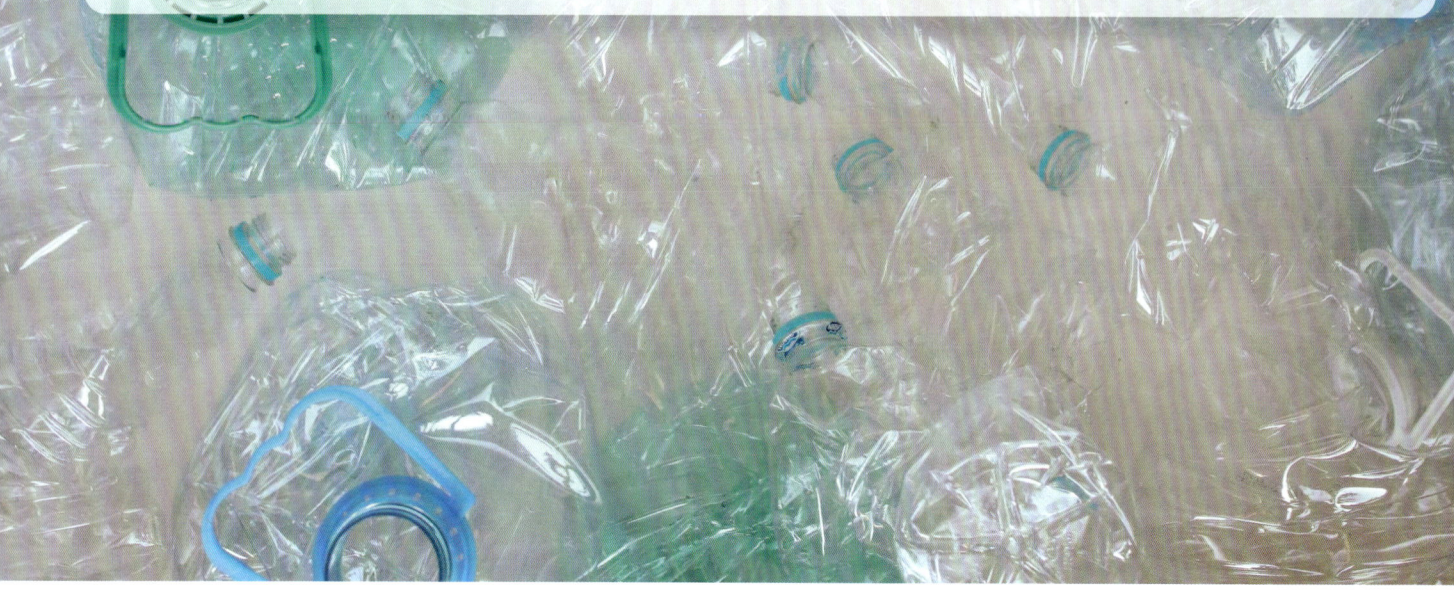

Seventeen projects selected

Funded through the Smart Sustainable Plastic Packaging (SSPP) challenge's future plastic packaging solutions round two competition, the projects will support a broad range of innovation, including:

- Encouraging consumers to move to refill and reuse
- New edible and biodegradable bio-based materials
- Advanced recycling technologies
- Plastic pollution mapping

The SSPP challenge represents the largest UK government investment into sustainable plastic packaging. The projects announced today will go on to support the achievement of the UK Plastics Pact.

Project categories

The projects fall into three categories.

Alternative materials

Five of the projects will be assessing the viability of a range of different plant-based biodegradable polymers to replace fossil-fuel-based plastics.

The projects will develop new, 'plastic-free' packaging for a range of food, personal and cleaning products, as well as more recyclable takeaway food packaging solutions. Through these developments, the projects have the potential to reduce the consumption of conventional plastic and cut down on difficult-to-recycle plastic waste that ends up in the bin.

Stimulating reuse and refill

Five other funding winners will explore how to stimulate more reuse, refill, and reduce single-use plastic packaging, both in our day-to-day grocery shopping and for food and drinks consumed 'on the go'.

They explore different aspects of the challenge, including consumer perception and behaviour, cleaning and hygiene, and logistics.

Increased recycling and plastic pollution mapping

Six projects focus on enhancing the UK's plastics recycling and stimulating the use of recycled plastics.

The innovations being explored include new sorting and recycling technology and novel digital approaches to packaging design and 'nudging' consumer recycling behaviour at home.

The final project will use satellite data and artificial intelligence to build a global plastic map to support the tracking and removal of marine plastic pollution.

Fit for a sustainable future

Paul Davidson, Challenge Director for the SSPP challenge, said:

'SSPP is working to make plastic packaging fit for a sustainable future, supporting over 70 research and innovation projects focused on consumer plastic packaging.

Taken together, these latest SSPP-funded projects offer up exciting opportunities to tackle plastic packaging waste holistically by reducing it at source, encouraging the rollout of reuse and refill business models, and driving more effective and sophisticated recycling and pollution monitoring and measuring.'

Project summaries

Algreen Ltd: bio-based multi-layer flexible pouch packaging

Algreen is a bio-based, biodegradable, and recyclable film that mimics the performance and aesthetic of single-use plastic film used for cosmetic, food, or fashion packaging.

The project will test the performance and durability of Algreen, assess the viability of at-scale manufacturing, and build the evidence base to support its launch onto the market.

The material's suitability for consumer goods packaging applications will be assessed and its environmental performance validated through a full life cycle assessment.

Blow Moulding Technologies Ltd: optimisation of plastic packaging through computer-aided design

This project will develop software to support more sustainable plastic packaging materials and designs for the plastic bottle industry. The plastic bottle industry has a global worth of £129 billion and is manufacturing bottles at a rate of one million per minute.

The software will automate the design process:

- Allowing packaging to be designed and manufactured with just the right amount of material
- Supporting the faster uptake of new materials and designs incorporating more recycled content.

Cauli Ltd: smart REusable cup dispensers and REturn kiosks

Cauli is building a smart reusable system that improves the user experience of borrowing and returning reusable cups. It aims to help the cafe industry transition to reuse systems by making the process of sourcing, returning, collecting and washing convenient and accessible.

Through an automated system that can be integrated to vendors' point of sale, users can borrow a reusable cup free of charge, and return at a smart return kiosk where the returned cups will be washed and redistributed.

Circular11 Ltd: turning low-grade plastic waste into timber substitutes

Circular11 has developed a process that creates construction materials out of film-based and cross-contaminated plastic streams that are not currently recycled. However, the process requires highly consistent feedstocks like PPE waste.

This project will develop a dynamic process control system that can expand the range of low-grade plastic feedstock materials that can be used while ensuring standardisation of product quality. This process control system could soon be licensed out to other manufacturers, allowing the extrusion sector as a whole to accept and recycle a wider range of mixed plastic streams.

CircuPlast: a green chemical recycling process for PET

The CircuPlast technology team from Stopford Ltd and the University of Birmingham will be adapting their chemical recycling platform technology to provide a sustainable solution for waste PET packaging.

CircuPlast uses hot compressed water as a green solvent to convert problematic waste plastics (such as pots, salad trays, and tubs) into high-value chemical compounds suitable for the circular manufacture of fresh PET plastic.

CLUBZERØ (CupClub Ltd): reusable IoT-enabled takeaway food containers, process and infrastructure

CLUBZERØ (CupClub Ltd) will be expanding its patented circular economy 'Sustainable-Packaging-as-a-Service' concept. This delivers, collects, washes, and returns IoT-enabled (digital ID) food and beverage containers to businesses and consumers at a competitive price compared to the single-use plastic alternative but with significantly reduced carbon dioxide emissions.

The aim of the project is to add a wider range of food container sizes and formats for different applications to the CLUBZERØ model.

Codipac: a hygienic, reusable food packaging solution

Codikoat Ltd will be trialling the use of their novel antimicrobial technology, Cydal, to reduce the environmental impact of reusable food packaging associated with the need for high temperature washing or chemical sterilisation.

Applied during the manufacturing process, Cydal kills viruses and bacteria. The project will develop and trial new reusable food packaging formats that contain Cydal and work with supermarkets and consumers to assess the viability of this alternative to single-use packaging.

Jara Partnership Ltd: eradicating unnecessary single-use plastic in personal care

With 50% of plastic bathroom packaging waste ending up in landfill or energy from waste, this project will see Jara Partnership develop a convenient refill system for personal care products.

Let's Go Zero Ltd: circular supermarket

weekly.shop is a circular supermarket concept that uses no single-use packaging.

ZEROWARE containers are reusable, made from 100% recycled materials and last for four years before they are recycled.

The system is designed to be frictionless for the customer, so that switching to reusables is easy. Customer orders are filled from local hubs on a flexible subscription; empties are collected when the next order is dropped off and brought back to the hub to be cleaned before being used again.

This project will support the final concept development for the service as a whole and container testing to verify the market proposition.

LitterLotto (partnered with RECOUP and Buckinghamshire Council): nudging behaviours towards consistent home recycling

The mobile 'LitterLotto' app incentivises people to dispose of litter appropriately through a combination of proprietary technology and behavioural science.

The 'Bin It to Win It' approach asks people to take a picture of litter as they dispose of it in exchange for the chance to win a prize.

This feasibility project will build extra app functionality, working with local councils to develop educational and incentive-based adaptations to encourage citizens to place used packaging in the correct recycling bins '@home'.

The app will focus heavily on single-tuse plastic packaging, while also providing residents with concise, simple information to help them recycle the right thing more often.

MarinaTex Ltd: in-flight biopolymer-based flexible film consumer packaging

MarinaTex will be developing a sustainable, marine-based, compostable 'polybag' (IFF CoPack) for consumer products in the travel industry that is compliant with industry requirements and waste facilities.

In collaboration with industry partners, the prototype bags, made from seaweed and seafood waste will be tailored to meet the travel sector's specific hygiene, transparency and protection needs and tested with users.

PlantSea Ltd: seaweed-made capsules for fluids and powders to replace polyvinyl alcohol (PVOH) film

PlantSea-Pack is a packaging solution that replaces single-use plastic sachets and bottles.

The seaweed-based films are regenerative and are not harmful to nature or aquatic life. The water-soluble film capsules can be used for liquid and powder concentrates for personal care products, such as shampoos, conditioners and creams.

This project will assess the feasibility and accelerate the scaling-up of the capsules and refill-at-home system for cosmetics, cleaning, laundry, food, or drink products.

As well as reducing packaging and eliminating the millions of non-recyclable plastic parts used in single-pack formats, it also reduces the transit weight of high-water content products.

Plastic-i Ltd: enabling the solutions to marine plastic pollution from orbit

Marine plastic pollution is one of today's most significant environmental challenges, with some 14 million tonnes of plastic waste flowing into the ocean each year and rising.

Using satellite data and artificial intelligence, Plastic-i is building a platform to facilitate the identification and removal of marine plastic on a global scale.

By detecting, mapping, and classifying floating debris, it will provide decision makers and clean-up operators with actionable insights to boost efficiency and measure the efficacy of interventions.

Recycleye Ltd: artificial intelligence-powered sorting for plastic packaging

Recycleye will develop its GRIP-R (Gripper Innovation for the Picking of Recyclables) concept. This is an artificial intelligence-powered (AI) plastic waste sorting solution with enhanced gripping function to handle the growing issue of contamination caused by films and flexible plastic packaging.

The project builds on the firm's innovative cost effective, AI-powered system and advanced robotics technology to design an even more effective sorting capability. This delivers high quality recycled plastic streams and increases the waste industry's ability to handle films and flexible packaging.

Sustainable Packaging Products Ltd: recyclable paper-based biodegradable frozen food packaging

Sustainable Packaging Products will be developing a fully recyclable, biodegradable paper-based packaging solution as an alternative to single-use plastic items for frozen food pouch and bag packaging.

TOPUP TRUCK Ltd: refill shopping on the doorstep

TOPUP TRUCK is a mobile zero waste shop that enables consumers to achieve a convenient transition to refill shopping by bringing the refill store to their door on an electric vehicle.

The accessible, fun, and community-centred approach overcomes some of the key barriers to refill adoption. This includes getting heavy containers to the shop, while continuing to offer consumers the option to buy as much as they need and to buy on impulse.

The project will partner with University of Arts London to conduct a study into how a mobile refill shop can be optimised in terms of customer experience and communications to widen refill adoption.

Xampla Ltd: plant protein paperboard coatings

Xampla will be developing and scaling up the manufacture of a plant protein-based coating for paperboard packaging applications.

Xampla's innovative materials offer a sustainable replacement to synthetic polymer coatings while maintaining excellent functional properties. Xampla's coating does not interfere with recycling waste streams and is compatible with home or industrial compost.

Applications for this technology include takeaway trays and boxes used in the food service market and wider paperboard fast-moving consumer goods and cosmetics packaging.

28 March 2023

The above information is reprinted with kind permission from UKRI Innovate UK
© 2024 UKRI

www.ukri.org

'Plastic-eating' enzymes to be deployed to combat waste polyester clothing

Scientists at the University of Portsmouth are to develop 'plastic-eating' enzymes that could help solve the ever-growing problem of waste polyester clothing.

Polyester is the most widely used clothing fibre in the world but is currently not a sustainable textile option and will likely end up discarded in landfill or polluting the environment. It is made from polyethylene terephthalate (PET), one of the most common consumer plastics.

Researchers at the University's Centre for Enzyme Innovation have already developed enzyme technology to reduce single-use plastics, including PET, to their chemical building blocks, leading to safe and energy efficient recycling. Now they have set their sights on creating a similar process for polyester textiles.

The process of recycling synthetic fabrics using enzymes will not be an easy one. It is estimated that these textiles account for 60% of clothes that are worn[1]

The addition of dyes and other chemical treatments make it even harder for these tough oil-based materials to be 'digested' in a natural process. Developing enzymes that can efficiently 'eat' polyester clothing, without energy intensive pre-treatment, is the biggest challenge.

Professor Andy Pickford, Director of the Centre for Enzyme Innovation at the University of Portsmouth said: 'We will develop enzymes that can deconstruct the PET in waste textiles, tolerating the challenges that this feedstock poses, namely its toughness and the presence of dyes and additives.

> **'We want a system that uses plastic in the same way we use glass or aluminium cans – infinitely recycled. The ultimate aim is to close the loop – however, this requires not only the technology but also the will to do so.'**
>
> – Professor Andy Pickford, Director of the Centre for Enzyme Innovation at the University of Portsmouth

'We will test the compatibility of our engineered enzymes with additives, dyes, and solvents to select those enzymes that are best suited to polyester textile deconstruction. Then we will apply these enzymes to appropriately pretreated waste polyester textiles in laboratory-scale bioreactors to evaluate the potential and limitations of scaling up the technology.'

Clothing has some of the lowest rates of recycling, with much of it being incinerated or ending up in landfill. While it is possible to turn good-quality oil-based textiles into carpets and other products, current recycling methods are highly energy intensive. Scientists hope that enzymes developed at the University of Portsmouth will help them create a environmentally friendly circular economy for plastic-based clothing.

Professor Pickford said: 'Our research will establish the feasibility of using enzymes to deconstruct the PET in waste textiles into a soup of simple building blocks for conversion back into new polyesters, thus reducing the need to produce virgin PET from fossil-fuel-based chemicals. This will enable a circular polyester textiles economy and ultimately reduce our dependence on taking oil and gas out of the ground.

'We want a system that uses plastic in the same way we use glass or aluminium cans – infinitely recycled. The ultimate aim is to close the loop – however, this requires not only the technology but also the will to do so.'

The research, which is funded by the Biotechnology and Biological Sciences Research Council (BBSRC), will start at the end of January 2023 and last for 18 months. The University team will work with project partners Biomimicry Institute, who will provide expertise in natural solutions to sustainability challenges, and Endura Sports clothing, who will share their knowledge of fabric dyes and provide samples of end-of-life polyester textiles.

6 February 2023

[1] (EEA (2021) Plastic in textiles: towards a circular economy for synthetic textiles in Europe), and are often chosen for durability.

The above information is reprinted with kind permission from University of Portsmouth.
© 2024 University of Portsmouth

www.port.ac.uk

A sustainable future: unlocking plastic recycling with table salt

Plastic recycling with table salt is poised to revolutionise the industry, offering a sustainable, cost-effective solution to enhance the recyclability of polyolefin polymers.

Researchers have discovered that table salt, sodium chloride, surpasses costly catalysts in enhancing chemical recycling for polyolefin polymers, constituting 60% of plastic waste. This breakthrough offers a safe, affordable, and reusable solution for improving plastics' recyclability, potentially transforming the industry.

A simple solution for plastic recycling with table salt

Muhammad Rabnawaz, an associate professor at Michigan State University's School of Packaging, emphasises the power of simplicity. His team's research, published in *Advanced Sustainable Systems*, highlights table salt's potential to outperform expensive materials for plastics recycling.

Despite plastic's recyclability claims, nearly 90% of plastic waste in the United States ends up in landfills or as environmental pollution due to the lack of value in recycled materials.

The team's projections indicate that table salt could reshape the economics of recycling, particularly in pyrolysis, a heat and chemistry-based recycling method.

In pyrolysis, plastics break down into gas, liquid oil, and solid wax. The wax component, often undesirable, can comprise more than half of the final product's weight. Traditional catalysts like platinum are effective but costly. Table salt, however, has proven to be a game-changer.

Table salt and its remarkable role

Rabnawaz's team found that table salt alone can eliminate wax byproducts during the pyrolysis of polyolefins. Using table salt as a catalyst, they primarily produced liquid oil with hydrocarbon molecules similar to diesel fuel. Additionally, the salt can be easily reclaimed with water, making it a sustainable choice.

Table salt also showed promise in recycling metallised plastic films commonly used in food packaging, such as potato chip bags. While it didn't outperform a platinum-alumina catalyst, the cost-effectiveness of salt makes it a compelling alternative.

A sustainable future for plastic recycling

Rabnawaz envisions a world where metallised films are no longer necessary, and his team is actively researching sustainable alternatives. They are also working to enhance the pyrolysis process to yield chemicals with more valuable applications than just fuel.

In summary, table salt's remarkable catalytic properties offer a simple, cost-effective solution to revolutionise plastic recycling, potentially diverting significant plastic waste from landfills and incinerators. The early success of this innovative approach suggests a brighter, more sustainable future for plastic recycling.

11 September 2023

The above information is reprinted with kind permission from Open Access Government.
© 2024 ADJACENT DIGITAL POLITICS LTD

www.openaccessgovernment.org

Chemical recycling: a game-changer for the plastic waste crisis?

By Becky Mckay

Chemical recycling is considered one of the most innovative solutions to our plastic pollution problem. Taking used plastic and chemically converting it into raw materials allows us to create entirely new plastic products without using fossil fuels.

But with an estimated 300 million tonnes of plastic waste being produced every year across the globe, is chemical recycling the key to the problem? Or is it simply another emissions-heavy process?

Some recent studies have raised concerns that this way of chemically recycling waste plastic may be more harmful to the environment than producing new (or 'virgin') plastics.

In this article, we'll look at the chemical recycling process, finding out what it involves and the many benefits and opportunities. We'll also ask, is this the solution our planet needs to combat plastic pollution, or is there more to be considered?

What is chemical recycling?

Chemical recycling is a waste management process that can convert various plastic (polymer) waste forms into their core components. It does this by transforming a plastic's chemical structure. This transformation produces materials suitable for manufacturing new plastics or other products.

Chemical recycling, as a term, includes several different technologies and processes. For example, some may use heat or chemical reactions to break down plastics. There are also differences in the type of product they create, which could be raw materials for making new plastics, fuel or other chemicals.

Below are three leading technologies used in chemical recycling.

Purification

This technique uses a solvent to dissolve plastics, which are then treated through a series of purification steps. The purification process can separate the polymer from any additives or contaminants, leaving just the material required for reuse.

This process can work in many ways, such as using different solvents. This means manufacturers can select the exact polymer or material they need from the plastics for their new product.

Depolymerization

This process, also known as 'chemolysis', converts a polymer into a molecular form known as a monomer. These molecules can be almost identical to those used in creating plastics, which means this process produces a high-quality material for developing new products.

However, the drawback to this process is that it can only be used on a specific type of plastic (for example, polyester). It cannot be used for most plastics in our waste system.

Thermal conversion

Thermal conversion uses heat to break polymers down into molecules. However, this technique produces fundamental chemicals that need further processing before they can be used to create new plastics. This greatly benefits some manufacturers as it provides more flexibility in their production.

While other technologies and techniques are being used, many processes are still in development, so there is hope these techniques will have more capabilities.

How is chemical recycling different from regular recycling?

The recycling process we're familiar with when our plastic waste is collected from our homes is known as 'mechanical recycling'.

In this process, plastic items will usually be sorted, washed, ground down (creating pellets) and reprocessed into something new. However, there are limitations to the quality of new materials that can be made this way, as well as a limited number of times the product can be recycled.

With chemical processes, however, these limitations are decreased. By taking a plastic back to its molecular level, there is, in theory, no end to the number of times it can be recycled. This is a great leap forward, considering the eco-friendly environmental impact of producing new plastics.

Many companies using chemical recycling techniques see this process as a way to 'close the loop' by creating an endless life cycle for waste plastics. The diagram on the next page demonstrates this theory.

The benefits of chemical recycling

With 60% of our plastic waste currently ending up in landfills or scattered across our natural environments, finding ways to tackle plastic pollution is essential to ensuring our planet can thrive.

Chemical recycling offers new, alternative options that can actively reduce waste, promote sustainability and eliminate our reliance on fossil fuels to produce plastics.

The benefits of this technique, however, are far-reaching. The following are some of the critical ways chemical recycling can make a difference.

Reducing plastic pollution

Plastic production has sharply increased since the 1960s, with an estimated 139 million tonnes of single-use plastic recorded in 2021. This amount of plastic leads to vast plastic pollution dumped in landfills and waterways.

One of the biggest dangers to this ongoing pollution is microplastics. As plastic products break down, they release small, incredibly harmful plastic fragments into the

environment. These microplastics threaten our wildlife and can also make their way into our food supply – meaning we ingest them, too!

By finding a sustainable way to eliminate this environmental waste and using chemical recycling methods, we can reduce the amount of harmful pollutants in our oceans and our food.

Lowering emissions

During their lifecycle, plastics emit 3.4% of global greenhouse gas emissions, with 90% coming directly from the production process and conversion of fossil fuels.

With natural gas and oil being used extensively to make plastic products and the massive demand for plastic worldwide, it's unsurprising that the emission levels would be so high. However, chemical recycling can help to reduce these alarming figures.

By reducing existing plastic back to reusable material, there will no longer be a need to use fossil fuels in production. This is a much more sustainable way to produce new products that drastically lower emissions, including CO_2 and other harmful gases.

Economic opportunities

With increased quantity and types of plastic that can now be recycled, businesses have more opportunities to benefit.

For example, with higher-quality recycled materials, companies can produce better products. They can also decrease their dependency on fossil fuels and lower their emissions from importing oil or gas.

On a broader scale, as plastic waste becomes a more valuable resource, this will reduce pollution and encourage technological advancements, creating more jobs and economic opportunities.

In the UK, the government is also encouraging the investment and development of chemical recycling in the plastics industry. They have committed £20 million to research, development, and deployment of new recycling technologies. This is part of the '25 Year Environment Plan', which aims to eliminate plastic waste by 2042.

In April 2022, they also introduced the 'Plastic Packaging Tax', which states that plastic packaging must be made from at least 30% recycled plastics. This is an excellent boost for the chemical recycling industry because it will drive more businesses to use the technology to develop their packaging.

While developments are still needed within the industry, it is predicted that with the right level of investment, up to one-third of all plastic production could consist of recycled materials within the next decade. Up to two-thirds of the growth in the plastics industry is driven by recycled materials.

The challenges of chemical recycling

Plastic production is one of the most polluting industries in the world and is a massive problem for the planet. Latest predictions suggest that by 2060, emissions from plastic are expected to reach 4.3 billion tonnes, which is a complex figure to comprehend.

Recycling products such as plastic seems like a great solution, but some statistics suggest that here in the UK, we're simply not doing enough of it. It's stated that in 2021, 2.5 million tonnes of plastic waste was produced, but only around 45% of that was recycled.

Despite the many benefits of chemical recycling, hurdles must be addressed before it can be fully utilised. Below are some of the challenges of chemical recycling.

Critical review and opinion

With technology in this field still evolving, the theory and practice of chemical recycling have come under critical review from environmental and scientific bodies.

A recent report showed that specific processes used in chemical recycling can release toxic substances – including benzene and lead.

The report also claims that, while technologies will become more streamlined, the entire process will result in a larger carbon footprint than traditional recycling methods.

Critics are concerned that attention should not be focused on recycling but on replacing single-use plastics, with further warnings that putting all efforts into chemical recycling as a solution would be a mistake.

Plastic industry leaders, however, dispute these findings and claim that chemical recycling can reclaim and reuse more plastic than any traditional means and provide employment opportunities across the globe.

Investment

Money and investment are crucial to transition towards a fully operational practice in the chemical recycling of plastics. Funding the development of new technology, building large waste and recycling processing facilities and obtaining high-quality plastic waste is expensive.

PlasticsEurope, representing manufacturers across the EU plastic industry, recently announced an investment into the sector of eight billion euros.

This vast sum is to be channelled into technology, policy and regulation. However, their official press release also acknowledges, '…this is just a starting point, and sizeable investments are still needed to capture the value of [chemical recycling] technology fully.

Investment is a big challenge for this industry. Various factors, including unstable energy costs, sourcing suitable plastic waste products and the need for ongoing technological advancements, make it even more challenging to source investment opportunities.

Environmental impact

Despite the many positive impacts of plastic recycling on the environment, in avoiding landfills and pollution, there remain concerns about the effects that chemical recycling will have on the environment.

As with any large-scale industrial process, significant amounts of energy will be needed, and large amounts of carbon dioxide will naturally be produced. These are a massive challenge for the industry because depleting natural resources and releasing harmful gases into the atmosphere directly contradicts the driving force behind recycling.

There are also concerns that many processes involved in chemical recycling use harmful chemicals and contaminants, producing hazardous waste gases and particles that can cause pollution.

If the industry is going to succeed, measures will need to be clearly outlined as to how they will combat these issues – in particular, how they will balance the benefits of recycling plastics with the impact of their reproduction.

Exploring a path forward for chemical plastic recycling

When considering chemical recycling as a solution to plastic waste problems, it's essential to consider the many aspects that drive the process. Breaking down plastics into something entirely recyclable to produce new plastics sounds like a beautiful solution.

However, when it comes to the actual environmental impact of this new technology, it will be critical that leaders in the industry take proactive steps to protect the environment. Whether through carbon emission reduction or management of hazardous by-products, many elements need to be fully regulated.

Without well-considered governance, there is a danger that what seems to be an innovative solution may just become another polluting problem.

With that said, however, any advancements which allow us to eliminate our reliance on fossil fuels are a step in the right direction. With further investment and more advanced technologies, there is hope that chemical recycling could begin a whole new manufacturing process.

Despite the challenges and critics' concerns, we know something must be done about our plastic problem. Chemical recycling offers a way to tackle that problem and lead us to a more sustainable, less polluted way of life. With so much potential, there's hope that this is one technology that will make a positive difference in our future.

19 September 2023

The above information is reprinted with kind permission from GreenMatch.
© 2024 GreenMatch

www.greenmatch.co.uk

Further Reading/ Useful Websites

Useful Websites

www.friendsoftheearth.uk

www.greenlivingblog.org.uk

www.greenmatch.co.uk

www.inews.co.uk

www.lifesabeach.org

www.mrcvs.co.uk

www.nationwidewasteservices.co.uk

www.nhm.ac.uk

www.openaccessgovernment.org

www.planetmark.com

www.plasticcollective.co

www.port.ac.uk

www.solentplastics.co.uk

www.telegraph.co.uk

www.theconversation.com

www.theecologist.org

www.theguardian.com

www.ukri.org

Further Reading

Page 37: European Environment Agency (EEA) (2021) Plastic in Textiles: Towards a Circular Economy for Synthetic Textiles in Europe. https://www.eea.europa.eu/themes/waste/resource-efficiency/plastic-in-textiles-towards-a

Glossary

Biodegradable waste
Materials that can be completely broken down naturally (e.g. by bacteria) in a reasonable amount of time. This includes organic materials such as food waste, paper waste and manure, which can be composted, as opposed to items such as plastic bottles that would take thousands of years to break down naturally.

Biodiversity
The number and variety of organisms found in a specific area. A balanced, healthy ecosystem will support a large number of species, making it rich in biodiversity. Human impact on the environment (for example pollution or deforestation) can reduce biodiversity, causing negative effects on the ecosystem.

Biotechnology
Biotechnology is the use of natural organisms and biological processes to change or manufacture products for human use. Biotechnology is widely used in modern society: for example, in agriculture, pharmaceuticals, the manufacturing industry, food production, and forensics.

Conservation
Safeguarding biodiversity; attempting to protect endangered species and their habitats from destruction.

Consumer
A consumer is anyone who purchases and uses goods and services.

E-waste
Electronic waste; discarded electrical items such as mobile phones and computers. There are strict EU regulations in place to ensure that e-waste is safely recycled or disposed of: however, the shipping of e-waste to developing countries is becoming an increasingly common problem.

Eco-friendly
Policies, procedures, laws, goods or services that have a minimal or reduced impact on the environment.

Fake fur (faux fur)
First introduced in 1929, fake fur mimics the appearance and feel of fur. Often made of plastic, fake fur can be detrimental to the environment.

Global warming
This refers to a rise in global average temperatures, caused by higher levels of greenhouse gases entering the atmosphere. Global warming is affecting the Earth in a number of ways, including melting the polar ice caps, which in turn is leading to rising sea levels.

Incineration
A method of disposing of waste by burning it into ashes. Incineration reduces the amount of waste that is sent to a landfill and can even convert waste into energy. However, there are concerns about the environmental impact of incinerators (air pollution, toxic waste, etc.).

Landfill
A type of waste disposal in which solid waste is buried underground, between layers of dirt. Biodegradable products will eventually break down and be absorbed into the soil: however, non-biodegradable products such as plastic carrier bags will not break down (or will do so very, very slowly).

Microplastics
Extremely small pieces of plastic (5mm or less), which come from plastic pollution in the environment. As plastic is broken down, small fragments break off and can be ingested by sea life, animals and humans. Some microplastics are purposely manufactured for use in cosmetics, such as microbeads. Many companies are phasing out the use of microbeads and replacing them with natural alternatives such as ground almonds or pumice.

Plastic
Plastics are a wide range of synthetic or semi-synthetic materials that use polymers as a main ingredient. Plastics can be moulded, extruded or pressed into solid objects of various shapes. This adaptability, plus a wide range of other properties, such as being lightweight, durable, flexible, and inexpensive to produce, has led to its widespread use.

Recycling
The process of turning waste into a new product. Recycling reduces the consumption of natural resources, saves energy and reduces the amount of waste sent to landfills.

Sustainability
Sustainability means living within the limits of the planet's resources to meet humanity's present-day needs without compromising those of future generations. Sustainable living should maintain a balanced and healthy environment.

Waste
Anything that is no longer of use and thrown away. Each year the UK generates approximately 290 million tonnes of waste, which has a damaging effect on the environment.

Index

A
actions, individual 12–13, 25, 29, 30–31
alternatives to plastic 6–7, 10–11, 15, 30–31 *see also* bioplastics
aluminium cans 7
animals 26–27, 28, 29

B
bags 6–7
benefits of using plastic 11
bioplastics 32–33, 35, 36
birds 26–27, 29
bottles 7, 11, 35
businesses
   plastic-free 15
   reducing use 25, 34–36
   single use plastics 9, 10–11

C
chemical recycling 39–41
CO_2 emissions 1, 23, 40–41
companies *see* businesses
Covid-19 26–27
cups 7, 35

D
decomposition 6–7, 32, 37

E
enzymes 37

F
face masks 26–27
food 12–13, 16–17

G
global plastic waste 2–4
grazing animals 28

H
health concerns 14, 16–17, 21, 22–24

I
incineration 5

L
landfills 2–3, 5
Lego 10

M
marine pollution 3, 18–25, 29, 36
media reporting 18–19
medical equipment 11
microplastics 1, 9, 14–17, 22–23

P
packaging 3, 11, 12, 15, 34–36
plastic-free, going 12–13, 15
policy, environmental 18–19

pollution 1, 3–4, 9, 18–25, 29, 36
production 1

R
recycling
   global rates 2–5
   problems 10–11
   technology 34–35, 37–41
   types of plastic 8
refills 12, 13, 34–36
Resin Identification Code (RIC) 8

S
salt 38
seabirds 20–21, 29
single use items 1, 3, 9–11, 12–13, 30
straws 11, 30

T
technological solutions 14–15, 19, 34–41
textiles 37
toothbrushes 7
types of plastic 8

W
waste management 2–5, 35–36
wet wipes 7
wildlife 22–24, 26–27, 29